Aideen's Promise
Allan Billard

Print ISBNs
Amazon print 9780228637158
Ingram Spark 9780228637165
Barnes & Noble 9780228637172
BWL Print 9780228637189

Copyright 2025 by Allan Billard
Editorial Supervisor JD Shipton
Editor Victoria Chatham
Cover artist Michelle Lee

All rights reserved. Without limiting the rights under copyright reserved above, no part of this publication may be reproduced, stored in or introduced into a retrieval system, or transmitted, in any form, or by any means (electronic, mechanical, photocopying, recording, or otherwise) without the prior written permission of both the copyright owner and the publisher of this book

Dedication

For Shirley and her quiet support that endured draft after draft.

Reviews

"With rich, vibrant language and evocative imagery, Allan transports the reader in time and place to the early days of the Grand Banks cod fishery, underlining the everyday struggles of ordinary people striving to survive and succeed."
Geoff Regan, Former Minister, Fisheries and Oceans Canada

"Allan's knowledge of the Atlantic fishery gives him a unique perspective in weaving a tale that bridges the Old World and the New. He had my attention from the first chapter."
Quinten Casey, Maritime historian and award-winning journalist

Acknowledgments

When I began writing the story of the first Europeans who crossed the ocean to what would later be called the New World, I found very few facts to go by. The Norse had come and gone with little trace, others returned many times for the fish but never chronicled their voyages for fear of being followed. So I was left with something W. O. Mitchell told me: "Just ask what your character would likely choose and let the story tell you where it goes from there." If I had a writer's muse, it would be Bill Mitchell, an outstanding Canadian writer.

The story did write itself but it was made decidedly more readable by the people who gave their time and expertise to review early manuscripts and respond with detailed assessments. Donna Jones Alward, Donna Barnett, Valery Compton and Tim Covell gave me advice on character development, pacing and much more. As well, Victoria Chatham and Tim Covell worked their magic as editors. To each of these friends I say, 'Thanks for helping me tell a better story.'

Table of Contents

Chapter 1 7
Chapter 2 16
Chapter 3 25
Chapter 4 38
Chapter 5 50
Chapter 6 70
Chapter 7 81
Chapter 8 96
Chapter 9 109
Chapter 10 119
Chapter 11 130
Chapter 12 141
Chapter 13 150
Chapter 14 165
Chapter 15 174
Chapter 16 193
Chapter 17 201
Chapter 18 217
Chapter 19 235
Chapter 20 246
Chapter 21 260

Chapter 22	269
Chapter 23	274
Chapter 24	281
Chapter 25	298
Chapter 26	305
Chapter 27	320
Chapter 28	328
Chapter 29	342
Chapter 30	349
Chapter 31	362
Chapter 32	370
Chapter 33	378

Chapter 1

Dingle, on the west coast of Ireland, May 1249

The dispatch rider didn't dismount. From his imposing height, he unrolled a scroll and in a strident tone, read its message, "His Majesty King Henry III of England and Lord of Ireland sends word of Vicar Maurice, his special envoy. Having left London four days ago, he is *en route* and will set foot here in Ireland's County Kerry on Thursday next. In accordance with his station, the vicar and his staff are to be domiciled at this, the only manor in the county. Make it ready."

The manor was a drafty tower, and the former occupant never cared to have his staff upgrade or maintain it. Despite its current state, Aideen, the mistress of the manor, assured the messenger the tower would be a pleasure for the new man to call home. For her, readying the tower meant securing a pair of ferrets to rid the place of rats, excavating the cesspool and swishing the bat droppings from ... well, everywhere.

On the day the new vicar was to arrive, she rose before the sun, pinned a fresh wimple over her thick auburn tresses and went down to the kitchen where her friend

Lil had already begun preparing the welcome feast. "It'll be poached salmon in a duck egg sauce," Lil announced with satisfaction. "And as my surprise for our new lordship, there'll be a fresh marzipan loaf to follow, drizzled all over with honey, of course."

Aideen agreed it all sounded quite appetizing and spent the rest of her day checking little details in all the corners on all five floors of the tower. She placed sprays of asters in goblets around the master's bedchamber and in the dining hall, dusted the broad sills below the double stained-glass windows. She paid particular attention to the ashes under the braziers and whisked away the residue until the slate beneath them shone. Three times that afternoon, she ran up to the parapets to watch for a sign of the new lord. The battlements reminded her of when archers once crouched there in defense of the residents against the Anglo-Norman invaders. The thought melted into a sad reality; *I fear the Norman overlords are now too well entrenched in our part of Ireland for us to dare shoot arrows at them.*

* * *

Out of the gathering dusk, a brace of riders crested the ridge above the tower. It was obvious who was in charge — the man in black mounted on a palfrey mare, a highly bred horse reserved for the high-born. A young page followed a few paces behind, looking uncomfortable on his donkey.

When he was but a stone's throw from the tower's substantial main door, the tall man stopped and doffed his broad-brimmed hat to survey the residence, revealing a flash of his bald head, a beak-like nose, and his ashen skin. Pulling off his riding gloves, finger by finger, he arched his back to study the drab walls dotted with ragged patches of moss and fern. Aideen noticed the man's formal cassock was rumpled and covered with dust.

The vicar adjusted his position on his mount, curled his upper lip and whined, "Humph. There's a sad spectacle. I don't see a great hall, or moat, or any outbuildings at all. That four-day ride from the coast made this saddle intolerable. My ass has never been so sore." He slid from his mount with a grunt and spat on the ground. "There had better be a hot bath waiting within and an ample bowl of herring bits."

One of the vicar's white silk gloves slipped unnoticed from his grasp and in the same instant was trampled into the sod by the twitchy palfrey. The young lad behind him gasped but spoke not a word.

Aideen raced around and down four flights of steps. By the time she reached the ground, stray wisps of her long hair had escaped her wimple. When she swung the heavy door open, Vicar Maurice was standing there, glowering.

With a sincere smile, she bade him enter. "Welcome to your new residence, m'lord."

* * *

Standing quite tall himself, the vicar's discomfort of facing a servant woman of the same height as him was obvious. To make the confrontation worse, she looked disheveled. For a long moment the two stood in silence, her smile and intent green eyes challenging his indignant scowl.

"I am Reverend Father Maurice, your new vicar. See to my mount and have a bowl of pickled herring brought to me in the bath, a piping hot bath, mind. While I'm soaking, have this cassock restored to a respectable state."

Aideen stood still, waiting for a polite salutation. The new vicar was unmoved.

"What? You might be an Irish peasant, but I have been advised you understand several languages, including concise directives delivered in your king's own tongue."

Not knowing what else to say, she replied in flawless English, "Is it *scadán* you'd be wanting, m'lord? It's the name by which we all know that particular species here in County Kerry."

Maurice jutted his lower jaw, scarlet seeped into his pale cheeks and his eyes slitted.

"*HERRR-RING*. And henceforth, you serfs will call the damned food by its proper name so we all understand. Now, in the king's own tongue, fetch me some herring bits and a flagon of warmed mead while you're at it." He barged by her into the vestibule.

For a moment she stared at the space where he had been standing, resolving to someday tell him the indifference shown by the Norman overlords to the population who spoke the Gaelic was galling. Then an urge struck her. She turned and in a voice a little too shrill, called after him. "Sir, I know you are to be the rightful lord of this manor, but it doesn't mean you're the second coming of Saint Patrick himself. Since you are now in Ireland, you should know that Saint Patrick called it what all the good people here call it. In the Gaelic, it's *scadán*, not herring."

Accompanying her independent streak, Aideen possessed the good sense to stop before she went too far, although she feared this was perhaps more than a step too far.

The new vicar cocked an eyebrow and flipped her a dismissive hand gesture. Aideen suspected he had perfected such reactions from years of coping with unenthusiastic servants. Hoping to punctuate her frustration, she stomped her foot on the slate floor. The thick rawhide sole of her boot made no sound.

To do the man's bidding, the mistress of the manor trudged down the narrow, winding steps leading to the strongroom, now used as a root cellar. She muttered to the cold, damp walls, "No wonder they call these high-born clerics the 'black robes'. It's more to do with their black *hearts*, I'll venture."

With a jar of pickled herring in one hand and a jug of honey mead in the other, she retraced her steps. Before she opened the door to the master's bed chamber, she pulled the cork out of the jug of mead to spit into it. When she marched across to the copper bath tub and held the tray out for the naked man, she pasted a smile on her face and said, "Your herring m'lord, along with a flagon of mead."

* * *

Aideen spent a sleepless night sharing Lil's straw mat on the kitchen's slate floor beside the door to the root cellar. She planned to rise well before dawn, but discovered Vicar Maurice had risen before her. "Summon the staff to assemble in the vestibule," he commanded, when they met in the kitchen.

With everyone in the tiny room, he clasped his hands behind his back and began. "As your current mistress has already learned, I am lord of this manor, plus all its livestock, including you serfs. Indeed, I have been granted leave to exercise the full governing authority of a bishop. As lord of this fief, you all live on these lands at my pleasure." He paused at that point and avoided eye contact before beginning again.

"Riding into your isolated hamlet yesterday, I noticed the village has but one public house on the high street. In the market square, there is little on offer and only a sprinkling of tradesmen's workshops. Without a daily market, where do the fishermen sell their catch? May the Lord forgive us, there's not as much as a properly consecrated cathedral. Nonetheless, if I ignore your God-forsaken dusty roads and abundant boglands, I see opportunity here."

After another pause, he waved one arm in the direction of Dingle and continued, "You do have a serviceable harbor. Scholars at both Oxford and the Université de Paris have studied the oceans and agree that the moderating temperatures of our generation have favored the sea, noting it has clearly warmed in this enduring period of most favorable weather. We must redouble our efforts to harvest and market the bounty coming from the sea around us. Of course, customs duties will be applied to all trade goods entering the port. A special levy will be collected on incoming loads of fish, as well. Should my fishermen land seafood caught by their own hands, only then will they be deemed exempt."

To Aideen, it was clear Vicar Maurice had rehearsed his presentation to the staff and planned to carry out his functions with a particular zeal the former master never mustered. "You watch," she whispered to Lil. "He'll be sure to slash the crofters' share of the bean harvests, not to mention add extra tolls on the farm animals sold locally, all to fill his own purse."

Vicar Maurice noticed Aideen whispering to the cook and glared straight at her. "As for you, I was informed you fell heir to a goodly number of arpents of your parents' peat bog along the shore at Beenbawn. Our Norman law stipulates you may retain your entitlement until the sea rises and erodes its toll from the shoreline, or until you stop paying me the rent of one shilling."

"I'll pay you no rent," Aideen blurted out. "My domestic services here over the past many years have always been considered fair exchange."

"As of today, your services are no longer required. You are dismissed from all duties. As vicar, village magistrate, and your landlord, I shall insist upon the one-shilling annual fee. Fortunate for you, you may cut and deliver bricks of dried peat as required to keep this drafty edifice warm."

He scanned the rest of the staff. "You all will now understand, living in a prosperous countryside is not without cost. There is one last thing. My recent experience with your Irish mead reminds me of drool. We shall bring in foreign-produced wines or sherry for my personal provisions. For all other incoming beverages, taxes shall be applied."

Aideen turned to her friend and said loudly. "There, you see. Our little community will become his private milk cow. Feather his own nest with your effort, he will. Next, he'll be expecting an elevation to sainthood."

She hugged Lil, stepped to the heavy oaken door, and walked out.

* * *

In her twenty-five years, she'd never pondered her future. Now she wondered if she had a future at all. Stomping her foot on a silk glove packed into the muddy path, she clenched her fists and promised to herself, "I'll make my own future and a better one for my people. There's nothing to be done about this troublesome priest, but I'll stand up for the good folks in Dingle — as soon as I find a piece of solid ground to stand upon."

Her own words struck her like a bad smell. She had no place to go. The only thing she could claim as hers was the vast peat bog near the harbor entrance that her father had left her. Without as much as a glance behind, she harnessed a mule and cart from the loafing shed and headed down the rutted track in the direction of Beenbawn.

Chapter 2

Beenbawn, at the harbor entrance

A low sea cliff jutted bravely out over the shoreline. Aideen walked to the brink and stopped to listen to the surf as it rolled billows of fluffy spume high onto the sandy beach. Two dirty-white gulls screeched as they wheeled overhead, hoping for a scrap or two from a small fishing boat entering the harbor. No other boats could be seen coming or going, no ocean traders slipped by on the pewter-gray surface that stretched to the distant horizon.

When a huge wave collapsed on the sand below, she leaned over the edge and a cloud of foam and spray wafted up to her face. Since her father first carried her high on his shoulders along these shores, she had always enjoyed the taste of salt as she counted the breakers crashing at the base of cliffs. "Never worry about the surf, my dear. It's the one power on our earth that takes away but seems to always give back." He had been older than most men when he married, but often claimed, "She keeps me young, so I'll always be watching my daughter grow."

Her mind drifted back to those carefree days when, if her father was busy with the peat moss, she roamed the shoreline and fragrant hayfields hand in hand with her cousin Bessie. As waifs, their world stood still while they chased butterflies and lost themselves in the blooming heather. The beach, the village common, and the farmers' fields provided no end of adventure for the two girls.

Aideen's father died when she was still young and her mother placed her into service at the tower house without delay. A little more than two years later, her mother died of a broken heart and overwork. Afterwards, Aideen rarely visited Beenbawn. When she was old enough to realize how important the family's peat bog was for the village however, she encouraged the neighbors to cut bricks of peat to heat their cottages for a small fee. Managing the enterprise reconnected her with the people of Dingle and gave her a small income.

It's fair for all, she decided. I'll save the coins in a stocking and only take a few back when I need a little sunshine on a cloudy day.

Bessie, too, had her childhood cut short when her parents noticed her interest in wild flowers and her remarkable talent for making her childhood garden flourish. She was sent to work for a crofter who needed help tending to his crops of barley and hops. After some years, she married the man and the two established a thriving brewhouse in the village.

Both girls followed different paths into adulthood, but their friendship endured, swapping gossip after mass, or sharing a sigh over what might have been.

*　*　*

When another wave crashed against the shore, Aideen's reverie was replaced by the memory of a comment Vicar Maurice made. His warning of rising sea levels and storm surges sounded ominous. Most people never heard the scholars' predictions of sea level rise, but rather than be threatened by the vicar's words, she remembered how her shore had grown and receded for as long as she had watched it. I'll make my future right here at the harbor entrance and I'll adapt, as the shore does, she vowed.

Leaning into the blustery wind with newfound determination, she raised her voice and announced, "Dingle is as near to the fishing grounds as any harbor in Ireland. I know plenty of folks who operate boats in and out of here, yet there's never been a market for their fish. If I were to build one, sure the fisherman of County Kerry will support me. And I'll support them. Together we'll build a substantial wharf and a fish house right here on this prominent spot. It'll be seen by foreign captains as well, so we'll attract their large vessels. In turn, we'll provide our goods in trade for their own."

The notion of recreating herself roused Aideen. Putting her hands on her hips, again she shouted her promise aloud. "When it comes to fish, I'll look for plump white codfish to grace the dinner bowls of the folks here in Dingle. I'll not deal in the Normans' herring though, for it's all oil and bones. I will do some trading in the foreign wine the vicar fancies and sherry too, whatever that is. Mind you, I'll not be paying any taxes if I do bring it in. These will be trades made for the people, not the vicar's purse, and by the God's bones, I'll do it right here under the new man's nose."

To build a structure sturdy enough to withstand the unpredictable storm surges, she knew she would need a pair of heavy-duty carpenters, men who could plant deep pilings and frame the decking to support a new storehouse.

Each week they are out here working, she resolved, I'll pay them a small wage from my savings and guarantee them free fish once we get underway. A few bricks of dried peat could warm up the deal, and maybe a half sack of Dingle beans from the vicar's root cellar will be on offer too.

The idea of pinching farm produce brought a playful smile to her face, particularly since the rising temperatures of the last generation brought bumper harvests across the land each fall. And she felt no guilt since she had worked side by side with the crofters to harvest the crop last autumn.

With a plan to complete a wharf and storehouse now fixed in her mind, she remembered a more immediate need. She still had no place to stay. Then a notion came to her. I'll reconnect with Cousin Bess. In return for serving the regulars at her brewhouse, she'll let me sleep in her loft.

Turning the cart for Bessie's, she cast a glance back at the cliff. In her mind's eye, the cove below was filled with a large foreign vessel, several small fishing boats and the sounds of brisk trading. To nobody in particular, she announced, "And out at the far end of the wharf, apart from the main building, I'll have my men fashion a plain shanty for me where I might rest my head at the end of each day. If Vicar Maurice's prediction of sea level rise does come to pass and it bubbles right up through the floor, well then it'll be the water for my weekly rinse-off."

* * *

Bessie was waiting on her stoop and greeted her cousin with a welcoming hug. "I expected to see you today. It's my loft you'll be looking for now. Am I right?"

Aideen rolled her eyes at how quickly news traveled through the village. "Yes you are and in exchange I'll be a good help to you around here."

Bessie was blunt. "You're always welcome and if the truth be told, I'll be glad of your help. Managing this place is more than a busy job for me and my man, plus brewing all the ale the folks here can drink.

You'll begin by scouring my cauldrons and the big brew kettle out back, so my mash tastes fresh roasted with each new batch." She listed off other kitchen duties before she mentioned helping to serve the patrons.

Aideen expected it, but as she entered the back kitchen and brewery shed, piled high with the previous day's pots that Bessie found too difficult to clean, Aideen felt lower than a scullery wench.

Determined to earn her keep, she knew from roaming the beaches how the aggressive surf would scour hard surfaces, so by using a seawater rub mixed with wood ash, she revived the scullery ware better than Bessie ever had. Nothing else renewed the bond between the two quite so quickly. Very soon after, Aideen was promoted to serving the customers.

When villagers stopped by to check on her, Aideen told them she was quite content. "It's different from what I was used to, but I enjoy listening to the regulars. There's one fellow, Paddy, he's a tree trunk of a man with a soft pink nose set atop a toothy grin. He peddles goods up along the coast of the entire county in a boat he made himself with ox hides. He even rides the River Shannon all the way up to Limerick on the large tides, he does. The best thing about Paddy, though, is that he always seems to have a mug full of gossip to share. There's no end to the giggles we have over his tales of what those city people are all about."

Helping at her cousin's brewhouse made her more determined than ever to get construction of her own enterprise underway. And the brewhouse turned out to be the best place to meet a pair of carpenters willing to undertake her project. She also met a miller who would deliver cartloads of lumber out to Beenbawn. Each agreed their efforts could be paid for with the package of goods she had suggested.

Late one night, in a steady spring rain, she set off with her cart to load out a few sacks of beans. The outside door to the root cellar was shorter than a standard entrance and ill-fitting. When Aideen tried to pry it open, it stuck at first, then creaked once too often on its rusted hinges. The noise stirred Lil from her light sleep, and with a cleaver in her hand, she slipped down the winding back steps to fend off whomever was taking their supplies. Finding that it was her best friend, Lil stayed to help Aideen load the goods she needed. "You'll be wanting this too," Lil said. She bent low to step through the doorway and spread a greased canvas over the payload to protect it from the rain.

*　*　*

At the day's first light, Aideen slogged along the muddy cart path and out to the shore. The usual Irish mist still blanketed the shoreline, but the salty air had the tang she enjoyed since the first time she strolled out along the cliff hand in hand with her father. The tide was as low as she had ever seen it and long brown fronds of seaweed swished around the timbers. Her two heavy-duty carpenters were already onsite, sloshing knee deep in the surf, inserting wooden stanchions between hefty boulders. Several stanchions stood braced and ready to be connected with fresh-hewn planks laid over the top. Aideen pulled off her wimple, corralled an armful of her woolen frock and hopped up on a pile of boards, tracking mud across the new decking.

"Fine work, men," she said as a curtain of hair cascaded across her face. "This mooring is as broad as I'll need. And sturdy enough too for any fishing boat there is, plus all the goods they're willing to bring me." She did a step dance and declared, "I am impressed how you made it extend out to where the water is deep. Now those deep-sea fishing vessels will find us a safe haven, the only one along this whole shore."

"What are you expecting in here, mistress?" one of the workmen teased, "the royal barge?"

"I hope not. The English king would not be welcome in this part of Ireland. Foreign merchants and Spanish fishing vessels do pass by out there, though. On occasion, we see Norse fishermen from the Faroe Islands. I'm hoping some of them will come ashore with their cargo. If the new vicar is correct, they may take away a load of Dingle's handcrafts too."

"Do you mean a windjammer as tall as the one there?" The other worker asked, stretching his arm and pointing beyond the mouth of Dingle Bay. Through the rising wisps of sea smoke, they could make out the silhouette of a sleek caravel in full sail. The ship was a fair piece out, but to Aideen, it looked to be heading straight in.

"Captain," she shouted and waved her wimple as high as she could. "We are right here, waiting for you at the entrance to Dingle's fine harbor. Come alongside, and if you have some fresh seafood in your hold, we would be pleased to make you an offer. Anything but scadán. This storehouse will serve any vessel that brings in the cod and salmon our people fancy, but as little as possible of the Normans' scadán." She kept waving and watching until the vessel grew smaller and smaller, not headed in at all.

"You might think me a fool right now, but there will be others just like them," she assured the carpenters, mostly to convince herself. "And when they do stop by, they'll learn how a regular visit here to offload their fish and buy our wares will benefit us all."

Chapter 3

Aarhus, Denmark

"Our church needs you, Niels, my boy. *I need you.*" Bishop Olsen reached up and gripped the lad's shoulder. As he spoke, his manicured fingers dug in deeply. "And think of your parents. How proud will they be when you enter the seminary?"

For Niels, a tall, tousle-haired boy growing up in Aarhus, Denmark's largest town, life had been easy. The idea of entering the priesthood had never crossed his mind, not once in eighteen years. But what the hand of a bishop compels, a mere altar boy cannot deny.

After a year of learning to read Latin and celebrate the church's liturgy, then another year studying the lives of the disciples and the works of church scholars, Niels was assigned to the bishop as his curate. "Our numbers are swelling too quickly for us to manage. I need you to ensure the faithful obey the many rules prescribed by Rome. The ones you should now know so well."

When Niels was ordained, Olsen gave him new instructions. "Go forth now and be a fisher of men and keep an ear out for local gossip. As a bishop, I need to know what goes on in the diocese. You're to be my eyes and ears"

Niels knew exactly where to start. The biggest news in town was the arrival of a whole crew of Vikings. They had contracted with the local boat yard to build them a new longboat.

Niels always considered Vikings to be a scourge, the worst test the church yet faced. He set off to get more facts from the master shipwright. *I must learn more of the rogues' intentions and prepare our defenses,* he pledged.

Approaching the gates, the sweet smell of hand-hewn oak timbers hung lightly in the air. Inside, wood chips flew in all directions. Niels could hear the pinging of rivet hammers as boat builders hurried to complete the large vessel. Seeing the priest arrive, the yard's master hurried over, bowed and tipped his felt cap.

"What brings you to our busy workplace, Father? Have you been wondering how my men and I are doing?"

"In truth, I was wondering *what* your men are doing." Niels pointed with grave concern to the tall longboat on a cradle at the water's edge. "That's a Viking's longboat. Are you supporting a pack of those despicable raiders in pursuit of ill-gotten booty?"

"Nay, nay, not at all. This vessel will soon be headed across to the far Norse colonies. My clients are from Norway and they may have once been thieves, but they plan to hunt walrus or trade farm implements to the settlers in Iceland who have already done the hunting for their ivory tusks."

The master shipwright explained it had been ordered, but not yet paid for. "The men want a wider, deeper vessel so they might load it with trade goods and travel the open ocean safely," he offered in his defense. "As I understand it, they will return and sell the ivory to carvers back here ... the carvers who furnish your church with prayer beads and statuettes. Wouldn't a reliable stream of rosaries be a good thing for your parish?" The master shipwright stood with a self-satisfied expression, expecting words of gratitude from the priest.

Niels was not sure any plans involving Vikings would be good for his parish. He wrung his hands in foreboding and set off for the cathedral. "I must speak with Bishop Olsen and seek his advice."

* * *

Niels found the bishop sound asleep in the chancery office, slouched in his favorite chair with his arms folded across his stomach. A missal lay open on his chest. The sound of Niel's hurried footfalls awakened the little man who looked up and cleared his throat. Niels didn't wait for permission to speak.

"There are Vikings in town, Your Grace. They're having a new longboat constructed for some special voyage. When it is finished, they plan to sail to the ice floes and secure a deal with the hunters of walrus ivory. When they do, they could wrestle control of the whole of the church's ivory supplies, driving the price up for us. And worse, imagine if they restrict the ivory for rosary beads and the many carved icons so important to our new parishioners? It's no better than raiding defenseless nunneries for their silverware."

"Yes, yes. The issue of walrus tusks is a concern and I do fear for the supply of ivory, Father, but of greater concern for us is the availability of fish and whale meat. Rome has issued a new list of Fast Days. Heaven knows the calendar grows ever more peppered with them. We all must choose fish on Fast Days, but our fishermen say their favored seafoods are becoming scarce. The church scholars tell us it's all due to rising sea temperatures — rising for a generation they have, and the fish have gone in search of cooler sea currents. If you couple this predicament with the many new converts the church is attracting, you will understand my real concern. What will the growing congregation eat if we cannot find the disappearing bounty from the sea?"

"Could this be connected to the pleasant weather the church scholars have recorded over the recent generation?" Niels asked.

"According to those scholars, yes. Warmer weather means warmer seawater. As a potential solution, I'm reminded of our

Norse sagas. They speak of abundant codfish beyond the ice-bound colonies you say these Vikings plan to visit. It is reported there is a place called Vinland with fish so numerous we all could be fed. Since these Vikings plan to visit the northern settlements in pursuit of ivory, I propose you confront them directly, and explain how our parishioners need seafood more than the church needs ivory. Tell them our treasury will pay a premium for a load of dried fish from those far-off lands."

"But Vikings are self-admitted thieves, Your Grace, not fishermen who can be trusted to respond to the needs of the church."

"Then you must do what it takes to convince them."

* * *

The master shipwright paced his boatyard as his craftsmen put finishing touches on the longboat, but froze in his leather boots when Arne and Stok Gunnarsson — the leaders of the Viking crew and the ones who commissioned the construction, appeared at the gates. Both brothers were burly, green-eyed brutes with straggly blond hair. They were, in fact, twins, but few would guess it due to one obvious difference. Stok lost his left leg below the knee after it got infected when a monk attacked him with a meat fork during a raid on an unexpectedly well-defended cloister in Ulster.

Arne led the way, wearing a frayed wool tunic with smatterings of someone else's blood. Stok hobbled along, dressed in clothes similar to his brother, but a step behind because his makeshift leg forced him to maneuver awkwardly around the construction debris.

Arne marched straight over to the longboat still sitting high and dry on the cradle. He snarled at the master shipwright. "Did you start working only when you saw us walking in? How long does it take you pigs' arses to finish banging together a longboat?" He wrapped his fingers around the handle of his ax and kicked at an ox hide pail of iron rivets, each one longer than a carpenter's middle finger and filed sharp. The pail's contents rained down on the workers.

"We've got better things to do than pick our snouts waiting for you to finish work. If all the glaciers are melted by the time you're done, we'll miss the walrus hunt. Then we'll have no tusks to sell to the church and no silver to pay for your work. Do you know what we'll do then? We'll row right back here and make you pick up every last one of those rivets and eat them." The carpenters recoiled in fear.

"No need to worry," explained the master shipwright. "We're waiting for the high water needed to launch. I expect it at midday, two days from today. And if you look over there," he pointed to an apprentice working next to the cradle. "The lad is building a sturdy dory to ride along with you on your important voyage."

Arne and Stok looked at each other. Neither could think how they might use the dory, but Stok elbowed his brother and pointed to the longboat's prow. "Look Arne, there! Those carvers are installing the dragon's head." To Arne, it did look quite menacing, and he relented.

"That dragon's head will warn each one of those walruses they should be terrified of who's aboard this new boat. Two more days is fair then, Master Shipwright. Your carpenters will be spared our sharp blades."

To celebrate Arne's change of mood, Stok suggested they head down to the docks. "There's sure to be several alehouses down at the harbor. We'll find some fresh ale and a pretty bar wench to keep us occupied for the two days until launch."

At the gate, Arne turned back for one last look at the longboat on the cradle, mumbling numbers. "I told them to make it wider than our old boat because I'm thinking of all the space we'll need to stash walrus tusks below the decks. Stok, how much profit do you think we'll make with this new one?"

Stok eagerly joined in. "That's easy. We'll collect all the tusks there are in return for a load of simple farming tools." His voice grew louder. "After we've captured the entire herd, the business of church ivory will be ours. Since we'll be the ones shearing the whole flock of sheep, we'll be the ones raising the price for the wool as high as we want."

*　*　*

Niels lingered outside the boatyard gates until the brothers walked out. "What is this talk of sheep shearing I'm hearing?" The question was delivered as a challenge. The brothers halted in their tracks.

"I am Father Niels — the one who guides the flocks in this parish. And if it's walrus you mean, not sheep, I am still the one to do the guiding."

Both brothers stood staring, without reaction. Niels faced Arne directly, thinking he looked the more powerful of the pair. Nervous, but striving to maintain an air of authority, he continued, "I was hoping to run into you men. I'm headed to the parsonage next to the new cathedral. Perhaps you two will join me?"

Arne scoffed at the priest's self-important manner but nodded. "I guess nothing happens in Aarhus without the church knowing, but since we do plan to sell to you, lead the way."

"I guess the bar wench will have to wait," Stok mumbled.

The trio walked side-by-side-by-side through the narrow alleyways. Local peasants were shuffled aside. At the parsonage stoop, Niels pushed the door open and invited the two to step in.

Arne waited for his eyes to grow accustomed to the dim, candle-lit antechamber. Still standing at the stoop, he grumbled to Stok, "This place stinks of incense and worse ... old men."

Niels darted around the room, snuffing out the smoldering incense. He grabbed his clothing off the one bench in the room and invited his guests to sit.

"There is a colossal task confronting our church, my men. We are bursting with new converts and Rome insists they all must respect meat-free days. Our problem is the supply of fish is unable to meet the congregation's need." Arne's expression revealed he had no idea why anybody would recognize 'meat-free' days.

Niels explained. "We Christians must choose fish or whale on Fast Days, but the fishermen say these days they see far less cod. Church scholars say the reason is the temperature. You are men of the sea. You must have noticed how the waters around us are warmer now."

Stok squirmed on the hard bench like a little boy. "How would we know? We stay on *top* of the water, not in it." He cackled at his own joke. "And last I checked, it's still awful cold."

"Yes, yes, it may be cold to the touch, but mark my words, the sea is growing less cold as time goes by. The codfish, we are told, have moved off."

Arne gave an impatient flick of his hand. "So, what's in this for us?"

"There is something for you, Arne. Indeed there is. I am told you plan to sail north in search of walrus. Our parishioners might welcome the ivory you seek to bring us, but they are desperate for fish. Now, my bishop has studied our earliest Norse sagas. They indicate there are great schools of codfish beyond the ice-bound settlements you plan to visit. Somewhere beyond Greenland there is a place called Vinland, with fish so numerous we might all be fed. I propose you go there with your new boat to find those great fish."

Arne flashed a not-so-subtle look of disinterest. Stok squirmed again on the hard bench and asked, "Why do you think we would want to go beyond Greenland?"

"If you retrieve a full load of dried fish for us from that far-off land, our parish will buy it from you … at a premium."

Stok turned to his brother. "Maybe the churchman has something here."

Arne glared at Stok. "I don't think so. No churchman is going to turn me into some wet-arse fisherman."

Niels guessed Arne hadn't planned to go as far as Greenland, but he needed these fearless seafarers and Bishop Olsen's dictate echoed in Niels' ear: *Then you must do what it takes to convince them.*

He also knew better than to think self-admitted sinners could be trusted to achieve a righteous end. However, there was nobody else capable of crossing to Vinland and returning with the bounty of an undiscovered fishery. He had to gamble.

"The master shipwright whom you have engaged is a faithful parishioner. He informs me you have yet to settle on a price for your new vessel. If you guarantee us three full loads of fish, the parish will advance the entire cost of outfitting your new vessel. Once you have completed your third voyage and Vinland's bounty fills our food bowls, your debt will be forgiven."

The offer was met with silence. Niels accepted it may have come as a complete surprise to the brothers, not to mention it was beyond his authority to propose, but he knew it was too valuable for them to ignore. He felt his heart pulsating as he waited. At last, he noticed a slight twinkle in Stok's eyes, followed by an ever-so-slight nod of Arne's head.

Niels drew a long breath before blurting out, "Fine. It's settled then. I was counting on you to become a part of the solution to our problem. Once you have opened the door to Vinland's bounty, nobody will go wanting during the forty days of Lent, nor the many other holy days."

* * *

The brothers stepped out into the alley, and Stok gripped his mouth with both hands, suppressing a belly laugh until he heard the parsonage door close behind them. At the click of the latch, he couldn't hold it back any longer. The snort that erupted from his nose covered the front of his blouse with yellowish phlegm, making him laugh harder.

"With our knowledge of the sea, making a profit with our free longboat will be easy." He elbowed Arne in the side.

"And what if there *are* huge schools of codfish beyond Greenland, like the man says? We'll trade a boat full of supplies to the settlers and load their fish aboard for the new converts to obey the meat-free rules. And don't forget, we'll still carry a load of ivory tusks under the deck."

* * *

Bishop Olsen paced the floor of the office. "I suppose in your mind you did the right thing, but God knows how long you will be gone. I'll miss you."

"Your Grace?"

"Well, Father, we can't have those men up and sail away without you. You'll be our assurance they will come back with a return on our investment. No need to prepare too much. I'm sure they'll share all their provisions onboard."

"I'm to go with the longboat to Vinland?"

"You made the agreement with the Viking brothers, and you will see it through to its end — the arrival of the goods, so as to speak. And who knows if there are peoples living in Vinland? *Skræling* is the name given to the heathens who are reported to live in the forest. You'll be able to confirm their existence and be the church's ambassador for them, delivering their first introduction to the Good Word."

"Are you officially assigning me this duty?"

"Think of it not as a duty, my son, but as a personal quest, your first opportunity to spread the Word to a new world."

Chapter 4

Aideen's new dock, Dingle

Morning after morning, before Bessie's brewhouse opened for the day, Aideen trekked through the drizzle to the work site on the shore. It seemed there was no end to the cartloads of materials, tools and refreshments the workmen asked her for. She could see progress each time, board by board, but more than once Aideen sat herself down on a pile of lumber and asked, "When will it be finished? All these hefty joists, the milled siding and these long deck timbers I'm forever delivering are forming up into a substantial warehouse on stilts, but it's taking an eternity, and my funds are becoming as scarce as the vicar's patience."

Her neighbors working in Beenbawn's now blossoming fields watched her making the frequent trips back and forth and christened her well-tramped cart path 'Fishwife's Way.' Fortunately for Aideen, it wasn't long before several of them took time from their crops to join her workers, volunteering whatever skills they possessed, be they swinging an adze or bending a back.

The sight of her storehouse growing day by day brought her some relief, but she had yet to see any foreign vessels. Doubt plagued her. "Will the efforts of so many ever produce even the minimum reward?"

As soon the structure was roof-tight, she scrubbed Bessie's brew kettle for the last time, collected her few belongings from her cousin's loft and led her mule cart back down 'Fishwife's Way' to fashion a home for herself. "The space at the outer end of my wharf is where I'll fall asleep from now on, comforted by the surf lapping against the pilings." She met the builders and asked, "Could you stay a day or two longer and knock together a tiny garret for me, perhaps with two openings on the corner facing out to sea? I want a westerly view to watch the setting of the sun and perchance to see a passing vessel or two." She imagined herself standing there, tall foreign caravels carefully beating to windward and one or two coming toward her to unload the fish she would market to the people of Dingle.

The first thing she did in her new space was spread a reed mat in one corner for her muddy boots. In another corner she set up a brazier. "At least the vicar can't take away the heat I get from the bricks gathered on my own father's land. And I'll charge him plenty for the bricks he needs to heat his drafty tower."

To add a final touch of herself to the space, Aideen stretched a colorful patchwork coverlet on the wall next to where her couch-bed would be.

It was given to me by my mother as a baby, but it's no touch of femininity for this place of serious business, she lied to herself. *It's there in case I need to run outside in a hurry to flag down the captain of an approaching vessel.*

* * *

Twice during her first night alone in the storehouse, Aideen jumped up from a restless sleep to check for passing boats. The first time, the moonless night was as black as her bricks of peat, and she chided herself for bothering. The next time she checked, it was still too dark to see anything, but she had been unable to go back to sleep since she first got up, so she stayed up waiting for daylight.

When the early morning sun did rise above the Dingle hills behind her, it illuminated the seascape more brilliantly than she had imagined. Slanting rays of sunlight beamed directly onto the sea surface in front of her, making the water twinkle.

"If I am ever to see a passing fisherman, this will be the time."

She scarcely moved from her vantage point until she did spot a vessel, then another. Neither noticed her new construction, though. Aideen realized the morning sun was shining directly into each of the captains' view. For any fishermen to notice her fine new warehouse at the mouth of Dingle Harbor, she would have to wait until after high noon when the sun crossed the sky to shine back on her shore.

After midday, a caravel did cross the horizon near her shore. Aideen untied her wimple as she had done before and waved it as enthusiastically as she could. "It sees us," she shouted to her workmen. "It's turning for here."

When the captain brought his vessel within hailing distance, he called out in Spanish, over the waves, *"Tiennes algún problema señora?"*

Aideen turned to her workmen. "It's not the Gaelic he's speaking, nor English, not even Norse. I could understand some of those. I have no idea what he is saying, so how should I respond?"

"He wanted to know why you waved at him in," chided one carpenter. "But since you don't speak his language, I doubt there is anything you could say to him anyway."

"You're right. I do hope he noticed this sturdy dockage you have prepared, though. Perhaps he'll spread the word of what we have here and, if he's not too mad, he might return when we are in business." She smiled as broadly as she could and waved the fisherman on by. "This reminds me I should get supplies ready to offer them, too."

* * *

Before the workmen finished, Aideen asked them to board up a rough part of the rafters for what she casually called 'drool'. She wasn't ready to explain what it was to be for or why she picked such an awkward location. She wasn't sure she would ever need it, either, but for the moment, it was enough to have a secure space set aside.

Late one afternoon, a weathered-looking vessel appeared. The man at the helm seemed to be maneuvering his big ship carefully, as if looking for the best channel into Dingle Harbor. When he got a clear view of the storehouse sitting on a substantial looking wharf he yelled to his crew. "Take us over there closer. We'll see if this new construction is stout enough for us to tie up."

When he noticed Aideen staring at him, he stood broomstick straight, tugged his auburn leather doublet tight to his waistline and edged his beret a little higher above his left brow, but carefully, so as to not expose the whisps of gray at his temples. As soon as he was close enough to be heard over the rolling surf, he called out, "*Kaixo andrea … Dia Dhuit.*"

The first part of his hail was unfamiliar, although to Aideen, it did sound similar to her name. She appreciated the second part of his phrase as she recognized the polite Gaelic greeting. He then continued in English.

"I am Captain Gabriel de Portu. This shaggy little fellow with me is Shicki, my First Mate. Our good ship, *Espiritu,* has been bobbing around out there in this dreary weather for a full month. Searching for whales, we are. Alas *andrea,* we still have no whales and our cider kegs are all but drained. We need to replenish our biscuits as well as find more to drink. What have you in the way of supplies for us to purchase?"

Aideen was fascinated by this mariner who spoke with such confidence, yet was so polite. Smoothing her frock and tossing back her curls, she hoped he might be a forerunner of a profitable offshore trade.

"Our first plan for this wharf is a place for fishermen to sell their haul," she called back, "but you are right. We do plan to bring in the provisions you foreigners require. Unfortunately, we are not yet ready to have a vessel tie up here. The channel into Dingle's waterfront dockage is nowhere deep enough for Spanish whalers, so you'll have to anchor out in the harbor and gather your supplies from the village by small boat. Be assured, however, when you next appear, we will have a sturdy dock ready for you to tie up and all the food and drink you need."

Aideen was pleased with his interest in her business and wanted to thank him, but she had one small concern. "One favor you could do for me, though, Captain, if I may? Twice, you called me Andrea. What you say does sound like my name, but if you please, call me Aideen."

With an affable chuckle, the captain called back. "Oh, forgive me. In my country, *andrea* is a polite greeting, the same as you say *A Chara*. Aideen it will be for you from now on. By the way, this coast has nowhere else with berthage built strong enough for us. If you plan to fill your warehouse with supplies, as you say, your new dock will then attract any vessel of my country's fleet. Perhaps we will become regular visitors."

"What I have in mind for this wharf may not fit with your endeavors." Aideen was reluctant to turn the handsome foreigner away, but she had no interest in the whaling trade. "To keep my promise to the people of Dingle, I'll be salting cod for them, not whales."

"Salt cod, you say. You salt your cod similar to what we do with whale meat?"

"Yes, we do. But if you salt whale meat, it still tastes strong. A codfish stays delicious, especially served with a mustard sauce or melted butter over it. In my mind, it's always mild, no matter how you prepare it."

"You might be right, but we are searching for whales, although these days they seem scarce hereabouts. Maybe they have already disappeared for the summer. If we don't locate one soon, we may be back for other fish to catch. If we do, we will be sure to call upon you."

Aideen was again encouraged by this polite foreigner. "Perhaps there are other products you could bring us from Spain? Soon I'll have plenty of room for whatever you might carry."

He glanced off in the direction of his home for a moment and said, "Well, since you are eager to become a merchant and offshore trader, let me suggest our distinctive wines. We Basques produce some fine varieties, much better than those from the rocky Spanish hillsides."

"You are Basque? Forgive me for thinking you hail from Spain. Wine would be welcome, but our former vicar once mentioned that Spanish wines are strong. They don't suit the Irish taste. Maybe Basque wines will be more to our liking."

"I can't say when it might be, but we will be by again at some point, and we could present you with some Basque rosé to try."

"One other thing, Captain, you said *Dia Dhuit*, a Gaelic greeting, when you first came close. May I say there are not many foreigners who speak any of the Gaelic?"

"Out here on the fishing grounds, we must keep good relations with the many nationalities we meet, including the Norse and the Irish. For the rest, we learn only the necessary words. If you are to trade in our wines, however, I will make certain I understand more of the right words for doing business with you."

Aideen remembered why she had her carpenters leave an unfinished storage area up in her rafters. "By the way, it would be best if you bring your wine in under the cover of night so as to avoid our new vicar's import tax. If you do, I will find the right words to thank you in *your* language."

"And I would be pleased to teach you, but now we must head to the village and get our supplies. I won't forget your interest in the fruits of our vineyards and your need to take delivery after dark to avoid the tax." Tugging his beret back down, he signaled to his First Mate to continue into the harbor. "I bid you *agur*. It means farewell, where I'm from."

"*Slán*. It's the way we say goodbye to friends here in Dingle."

* * *

After a night of more self-doubt and only the undertaking of a well-mannered Basque sea captain to deliver wine sometime in the future, Aideen went back to Bessie's brewhouse for a hearty breakfast. She spied Paddy, the riverboat man, eating his first meal of the day at his usual bench, pottage on a trencher and swishing it down with a mug of cloudy ale.

"Could you take me to Limerick? I hope to meet a Norseman, maybe at the market there."

Paddy took another swig from his mug, wiped his chin with his sleeve and furrowed his brow. "Mistress Aideen, I'm always up for a merry chase, and Limerick does draw men from near and far, but why strike up with a Norseman? Are there no worthy lads for you in Dingle?" He tugged his tunic tight over his protruding belly. "Like right here."

"I need no man-friend. I need to contact a Norseman for information he might have on fishermen in the Faroe Islands. It's a load of their cod I need."

"Sure, a plate of poached cod might be a sweet meal, but why would you want to go all the way to Limerick. We have fine fishermen here, and the sea has plenty of other fish."

"I don't want the bony *scadán* our fishermen bring in, and no Kerryman wants to go out far enough to find me a cod." Aideen's frustration showed. "Since I learned a little Norse from my dad, I hope to contact the Norsemen who'll come in close here and transfer their load to our Kerrymen so they can bring it ashore without paying the tax."

"You plan to deal with the Norse fishermen and outwit our Norman vicar? You have the nerve of the Wee Folk. You'll need a pot full of their special luck, too."

"Are you saying you'll help me?"

"I'm saying the next flood tide will be running upriver in five days. We'll be riding it all night. Bye the bye, dearie, have you noticed how the tide seems to be out right here?" The cocksure boatman rapped his mug on the table and smiled. Aideen refilled

Paddy's mug and thanked him for agreeing to help.

After the last of the evening patrons staggered out the door, she grabbed her shawl and led her mule back to the tower house. At the cellar door, Lil helped her carry out the three more sacks of field beans she would need for the trip to Limerick market.

* * *

In the fitful hour just before she awoke, Aideen saw brown fronds of kelp wrapped around the wharf pilings. Their waving motion hypnotized her until the long brown seaweed transformed into fish, so many that she couldn't see the seafloor for the circling masses. Reaching in to grab one, it slithered from her grasp, and she woke.

She sat up on her cot and asked, "In my dream, I saw all the fish I needed, but I couldn't hold onto one. Is this a sign? Will I never get my fish?"

She had no time to answer her own question as a horse whinnied outside. Darting to the wall for her shawl, she looked outside. Vicar Maurice was sitting on his palfrey, nosing around the storehouse. He directed his mount toward the back of the building, where she knew he would spy the shelter with the mule and cart she had taken weeks before. *I hope he doesn't see his three sacks of beans,* she thought, and hurried outside.

"You should know I am aware of your actions and how you continue to use my cart and beast to do your personal business."

"I need the cart and mule to deliver the bricks of peat to warm your home. It's the one source of income you've left me with."

"Just so. I'll not be unreasonable. If or when you ever do any real business, in the fish I mean, you'll need it on Fridays to deliver my requirement of herring bits."

He reined his mount around and trotted down the path.

Aideen shut the door to her garret and flopped down on her cot. Then she remembered the three sacks she had stored in the back. She hopped up and checked if she had hidden them well enough. She had. *In this game,* she thought, *there are still more ways to pluck the goose.*

Chapter 5

Aarhus Boatyard, Denmark

It was launch day. The master shipwright cocked his new felt cap a little off center and opened the gates to welcome the families of his workforce for a celebration. The air was bathed in the aroma of honey mead as workmen gathered in small groups, enjoying the free-flowing drink. Wives and sweethearts chatted while their children tossed balls to the dogs running wild, each one yapping for attention. A jester in multi-colored regalia added to the merriment, dancing and entertaining anybody who would toss a penny into his cockscomb cap.

Stok sensed the excitement as soon as he hobbled through the gate. He spied the master shipwright and called out, "Hey, shipwright. We spoke to the priest. He wants to be our new partner. Good news for you as he's going to cover your full costs." A flush of relief showed on the man's face.

"We'll have four of these round puncheons filled with mead and carried aboard for her maiden voyage then," he said. "And Stok, we had the craftsman who carved the dragon figurehead whittle you a replacement leg from a sturdy ash tree. We believe you'll need it for wrestling the walrus."

A young apprentice with wood shavings tangled in his curls marched shyly up to Stok to present the new peg leg. Stok immediately flumped down on the grass, fixed the handcrafted limb onto his stump and stood with a flourish. He flipped a coin to the apprentice, then stomped around in a big circle, bellowing, "Look at this, Arne. It fits much better than the old one, and it's decorated with a pair of mermaids with great big breasts being chased by a sea serpent."

Arne scoffed. "You remind me of a one-legged rooster chasing its tail feathers, but you do get around better now."

Stok kept stomping in circles until the whole crowd was staring at him. When he stood still, he winked at a group of wives and galumphed across the yard, this time backward. The wives applauded.

"I'll be sad to leave this audience behind," Stok said.

* * *

At high noon, Bishop Olsen led a procession onto the site. His sash of purple silk glistened in the brilliant sun. Trying to avoid the scruffy mutts and the children's balls they were chasing, Olsen headed straight for the cradle and the new vessel perched high upon it. "Such a fine boat you've crafted, Master Shipwright. A true match for its fine mission. The christening ceremony will begin." He turned to Arne. "Naturally, we'll be pleased to extend the blessing to your entire voyage." Without waiting for permission, the bishop encouraged the crowd to gather in an arc at the top of the slipway.

The crowd was silent while the bishop dipped his silver sprinkler into a container of holy water, waggled it forward, then aft. Arne stood still, staring at his boots.

"Blessed be this vessel. We ask you to protect it from the perils of the deep, its master Arne Gunnarsson and those who labor upon her deck. Bring them all home to the safety of your harbor of peace as they carry with them the bounty of Vinland to feed your servants here in Aarhus."

As soon as the blessing finished, two workers climbed aboard and gave the signal for seal oil to be slathered on the slipway rails. Two others hammered the chocks out, and the longboat slid down the incline. When it splashed into the estuary of the Aarhus River, a thunderous cheer erupted across the yard. Moments later, the crowd gasped as the longboat leaned alarmingly to one side.

"Not to worry. It must be the extra kegs of mead," assured the master shipwright. "They must have placed them all on one side." The workmen hurried onboard to roll two of the large kegs over to the other rail, and the longboat settled proudly, steady on its keel.

* * *

More honey mead was passed around, and several of the carpenters treated their youngsters to a mugful. The lads' mothers grumbled their annoyance. Men with spools of light hemp twine, bone needles and palm pads climbed aboard to fit the broad, striped awning over the rear deck. Next, with longer needles drawing twine coated with beeswax, they sewed the broad sailcloth along the length of the cross spar.

Once rigged, the spar was hoisted high up the mast and secured with rope and shackles. As if in appreciation, the square sail of tarred reddish wool fluttered in the breeze, then billowed out. The longboat flinched like an anxious stallion, and the crowd cheered again. Laborers carted more provisions aboard, along with tar buckets and armloads of iron farming tools to be traded in the distant settlements. The young apprentice led two unhappy goats up onto the deck, latching them inside a pen he had fashioned in the longboat's narrow forepeak.

After the hubbub subsided, Niels stepped forward with his few personal belongings, plus a portable mass kit containing the items he would need for conducting remote services. It all fit into a new satchel of calfskin along with a notebook and manual.

Olsen put his hand on the priest's shoulder. "You may be ill at first, but you will learn to be at home on the waves soon enough, and there'll be no need for turning back."

"God willing," admitted Niels. "I do enjoy a meal of fish, but I can't say the same for the open water they live in. Nonetheless, I accept my task and will carry the Holy Word to any of God's children I encounter over in the land of vines. While doing so, I will ensure the load of codfish is secured."

"The blessings of our Savior be upon you." Bishop Olsen delivered the words with a self-assured smile.

"Indeed, may they be upon us all," Niels responded, and bowed deeply enough to cause the noonday sun to reflect off his freshly polished tonsure.

Niels had grown proud of the trappings of a priest, particularly the one badge of his new position he delighted in most — his shiny, smooth tonsure. It was the mark of a young man who had succeeded. "I'll miss all the fuzzy brown curls I had up there. It would be useful now as I head out onto the cold ocean."

* * *

"Climb aboard, priest," Stok bellowed. "I doubt you'll be into the extra mead the boatyard has provided. My bet is you won't be eating or drinking at all until you get your sea legs."

"Could you direct me to my berth?" Niels asked.

"We have no sleeping berths," Stok said. "Stash your mat and that handsome satchel of yours up next to the goats. Come nightfall, spread out on the deck and snug right up tight to the critters. They'll keep you cozy."

Stok watched as Bishop Olsen reached into a pocket hidden in the folds of his robes. "Father Niels, I have a parting gift for you." The bishop held up a silver chain bearing a pendant. Niels noted the image of Saint Christopher on it, but the bishop turned the pendant over to show another image on the reverse side. "This an engraving of my dearest friend, Bruno, the Prince Bishop of Bavaria. He has been minting these unique medallions at his private refinery. We save a few for special occasions such as this. Pure silver medallions are rare enough, but one with the image of a saint is a treasure indeed." Olsen droned on, "The scholars say if a traveler is in distress on his voyage, he should caress this medallion, then continue, comforted by the knowledge he is in God's company. I trust it will be of service while you are on this righteous mission."

He stretched an arm up to the shoulders of his subordinate as Niels bent low. Olsen draped the chain and religious icon around his neck, and Niels kissed the man's polished ring before turning to climb aboard.

Stok's gaze fixed on the glistening medallion. Arne caught the flash of light reflecting off the bishop's weighty bauble.

"Do you see his ring?" Arne whispered to Stok. As soon as the settlers in the colonies see all the iron tools we have for them, plus these two animals here, I'll trade my share of the silver they give us for a gold ring as big as his. And mine will be even shinier than his."

* * *

Arne scrutinized his new longboat from end to end. Satisfied it was ready, he signaled to his eager crew to climb aboard, settle onto their benches and be ready to insert their oars through the oar ports. Each oar was three times as long as the rower was tall, but carved from slender, young pine and surprisingly light.

Arne only picked crewmen who felt at home on the sea, and he had no patience for any man who refused to take the same risks he took. As he surveyed the two lines of rowers in front of him, he knew he had chosen his crew wisely, except for one figure who was definitely out of place. The tall priest had taken up a position beside the imposing dragonhead. Dressed all in black, with his arms spread wide, Niels was performing deep breathing exercises. Arne shook his head and called out, "Bend your backs, men. Next stop is the Faroe Islands, four days away."

The colorful felted sailcloth billowed out again as the longboat headed into the wide Bay of Aarhus. Stok took the tiller and steered to the northeast, while Arne arranged his wooden compass and crystal in a niche under the awning, confident that he would need only these two navigation tools for the entire crossing.

"Those are storm clouds up there, Arne," Stok warned. "I'm betting there's an overnight blow coming in."

Arne knew his brother was often right when he predicted the weather, but it was too late to prepare now. "It's too soon yet this spring for major cyclones," was his rebuttal. "Bad weather comes in from the southwest and later in the year."

By late afternoon, the longboat was well up the straight between Denmark and Sweden. A strong current flowing up from the south pushed them along, but the wind was whipping down from the north. The competing forces created short, choppy waves. The crew wasn't yet accustomed to their new vessel's reaction to the sea conditions, and none of the rowers could sustain a steady rhythm with their oars. At one instant, the longboat rode high atop wave crests, followed by headlong slides into troughs.

A huge black wave hit them broadside, sweeping two kegs of fresh mead across the deck and striking a rower named Olaf in the chest. When he fell to the deck, the kegs rolled right over him to the far rail. Arne joined Stok, gripping the helm as much to keep from following the mead overboard as to keep the longboat pointed into the wind. He called for the spar to be lowered to half with the sail furled tightly to it. Arne mumbled a few genuine words of praise for the carpenters back in Aarhus when a series of low groans came from the longboat's hull as it met the repetitive pounding of the waves.

Unable to maintain his footing and ignored by the crew, Niels opted to lie on the deck, his knees tucked into his chest and his eyes shut tight. He waved one hand searching blindly for something to hold on to, and with the other, he rubbed and rubbed his Saint Christopher's medallion, hoping it might relieve the horrible snarls in his gut. It became his 'worry stone' although it relieved nothing. He examined his conscience over and over, but couldn't think of any sin he ever committed, or imagined committing, for which such penance must be extracted.

Nobody slept or ate that night.

* * *

The sky brightened with the dawn on the third day, but the ocean's surface remained a series of long rolling swells. Arne took a compass bearing and swung the tiller so the boat pointed to what he sensed was the correct heading for the Faroe Islands. Olaf, the oarsman who had been hit by the mead kegs, cursed at Arne. "I knew this new boat was a mistake. It's too wide to take these seas as well as our last boat did. You'll have a lot more trouble with the helm now, Arne. Mark what I'm telling you."

Olaf could offer an opinion on just about any topic, for sure anything to do with boats and the sea. He was as strong as any man aboard, and Arne depended upon him to set the stroke rate and more, but the man's bloated gut hung well over his breeches and slowed his gait. Nobody had ever seen him wash, and a fusty odor clung to his wool tunic. His long, downturned nose resembled an owl's beak, and to mock him, the others gave him the nickname *Olaf the Ugle*, which meant 'The Owl.'

Arne knew Olaf was often right, but in his own defense, he declared, "The Faroes aren't much farther now. The sea will calm out from here on."

After another night of fair winds, Arne pointed out a series of rocky islands with one well-ordered settlement, much the same as all Norse outposts. Tired, but relieved at the sight of a well-equipped boat basin, he guided the longboat between two rocky outcrops and straight up to Tórshavn's broad wharf.

Except for small boats and a pair of scrawny cats, the waterfront was abandoned until a trio of inquisitive youngsters hustled down to take the longboat's lines, hoping to earn a pence or two for their effort. A few more residents gathered to scrutinize this unexpected arrival, hoping there might be interesting supplies onboard.

"Well done, my lads." Stok announced, tossing a coin to each of the boys. He scanned the growing crowd and spotted a maid in a faded pink-dyed jumper pinned at the neck with a bronze brooch. He elbowed his brother. "See the wench over there. See her big pin? I had real trouble unpinning the clasp on it when we landed here before. You remember — we drank all day at the little inn." He pointed down the winding track to a building looking exactly the same as each of the others, except for an oversized wooden mug hanging from the second story. "Arne, remember the tavern wench in there? She was tall, blonde, and quick with the drink."

Arne groaned. "There are already more men in these settlements than women. Don't be starting one of your brawls over a barmaid. All we need to do is unload this farm gear, load their stash of tusks back aboard, and see if they have any fish for the priest."

* * *

Meanwhile, Niels forgot the agony of the voyage from Aarhus. The clenched muscles in his stomach loosened and he no longer wanted to die. One thing he was grateful for was that the rain had flushed the vomit off his robe, leaving the familiar stench of wet wool. He put his Saint Christopher's medallion back around his neck and stepped down onto the pier. "Solid ground," he said. "God be thanked."

Wading right into the growing crowd he engaged with anybody who would listen. "I've heard rumors of visiting Celtic friars who settled on these islands, generations back. Can anyone tell me how I might investigate any of their sites?" The trio of boys who secured the longboat's lines offered to show Niels an ancient cave they had 'discovered'. The eager priest followed them up a series of low hummocks where they claimed old friars once dwelled in a grotto.

Twenty minutes after Niels scurried up the steep cliffs above the village, the boys dropped into a cave they called their lair and pointed to artifacts scattered carelessly in the shadows. Niels was overwhelmed by the religious significance of the treasures and squatted to make notes on each.

"Look at these old wooden crosses. They must be Celtic. Those first Irish friars surely carved them. They prove Christians did toil here, I mean holy men and monks, as told in the sagas. They practiced our faith on these remote islands and left these pieces here as a sign of their beliefs, maybe for newcomers to use."

* * *

One burly man stood at the back of the gathering crowd, looking distrustful. Arne grabbed his brother's sleeve. "Stok, see the sour-looking Viking over there, the one with the tattoos on his neck and the fishermen's blade tucked into his long walrus hide boot, isn't that Baldr? I can't say he was ever a friend of ours, but he was no enemy either. He could be willing to help us trade this ironware for walrus tusks. Have the men unload the tools."

Arne climbed down to the wharf, headed directly over to the raider-turned-fisherman and asked him about tusks. A sarcastic smile spread across Baldr's face. "Yeah, I remember you two, but you're too late. With all the sea squalls around here this spring, the walrus moved off. You'll have to go on to Iceland or maybe farther to find the big herds. We may be able to trade you the few tusks the local carvers have stored up, but no more."

Arne tried again. "By the look of you, you must be a fisherman now Baldr, so you have dried cod for us, right? We'll accept a full load of your best dried cod for these supplies here."

"Not possible," came the reply from another fisherman. He was shorter, shaved clean, with his arms covered in tattoos. By the look of him, he had also been an active Viking. "You have to understand how we live here. We prefer fresh seafood, not dried fish. And with all the recent storms, there's no fresh cod either. All we have right now is salmon."

"What about the dried fish left over from last winter?" Arne persisted.

Baldr jumped back into the conversation and tried to be positive. "You're right. We did put up dried fish last fall, but we won't have enough to trade until later this season."

"This isn't the first visit Stok and I have made here. We know of a dozen other villages on these islands. We'll do a deal with them for tusks."

"Try those islands if you want," Baldr suggested, "the folks there will tell you the same thing. Walrus don't come around here much anymore. We do hook a few fish, but the merchants' guild in Bergen has first call on it all, including the ivory and any other product we have coming and going. They control us. And they've banned us from trading with outsiders."

* * *

Half of the load of farm implements had already been heaped on the wharf, but the locals only shouted out low offers. Arne sensed further negotiation would produce nothing and ordered the crew to stop unloading.

"So, Baldr, in return for leaving the supplies we've already unloaded, we'll take a credit agreement from you. You and I will agree on their value, minus any walrus tusks you have for us. We'll return when you have dried fish, and nobody has to tell the merchants in Norway anything." Baldr nodded his agreement and led the crew down the lane to the inn.

Arne moved next to Baldr. He gulped his ale and began scratching detailed markings into a tally stick representing the agreed value of the iron wares. He broke the stick into two parts, gave one to Baldr and explained, "We'll be back and join these two parts back up to remind us what our deal was. If the two parts still match, you give us the value in fish."

Baldr took his half of the tally stick and handed a sack to Arne. "These few tusks are all we have here," he said, still sounding wary. The butt of a long white tusk poked out of the top.

"Look at this," Stok murmured with a grin and grabbed the tusk like it was a club. "This is the size of my arm."

"Alright," agreed Arne, wrestling the tusk out of his brother's hands and fitting it back into the sack. "I guess these can be sold to some carvers, but Stok and I will return later this season. We'll match the two halves of the tally stick, and you'll pay us what it says."

"That's agreeable to us."

After another hour of what Stok hoped would be a full day of drinking, Arne stood up. "We're leaving. Stow these tusks below deck and go fetch our passenger back from wherever he went."

It took Stok a little more time and a lot more effort to climb the same terrain Niels had eagerly scampered over. When he arrived at the grotto, he was short of breath and shorter on patience. "Priest, we're leaving right now. Pack up your trinkets. We're taking them on a long voyage."

* * *

After three more days and nights battling rough seas, but with the wind at their backs, Arne sighted the craggy, volcanic landscape of Iceland. The snow-capped glaciers glistened in the brightening sky. The crew rowed with more enthusiasm when they spied the entrance to the ice-free harbor. Several residents had noticed them from a distance and waited on the wharf, curious to see what this unfamiliar transport brought for them.

"This is a better reception than we got from the crowd in the Faroe Islands," Arne noted and directed the crew to spread a few samples of their cargo on the dock, plus the two goats.

The Icelanders gathered to examine the supplies closely, but before allowing bargaining to begin, Arne waded into the crowd and laid out his conditions: "We'll give you all these provisions for free in return for twenty pairs of tusks and a load of dried fish."

"That doesn't sound free to me," complained an Islander, "and we don't have any ivory for you." The response was the same from the others. "The walrus have all moved to where there's still ice, maybe up the other end of the island."

"There's supposed to be enough walrus all around here to keep the entire world supplied with ivory," Arne growled. "How far do those fat arses move?"

The crowd responded with the same excuses Arne heard in the Faroe Islands. One lady, who spoke with the conviction of an experienced hunter, explained, "They move with the ice floes. You'll have to go another hundred leagues to the other side of Greenland now. There might still be ice cover there."

The last comment discouraged Arne. "Why do they call this place Iceland if there's no ice?" A few snickers echoed back to him, but no explanation.

"And as for dried fish, we won't be putting any up until later in the season." To make her point, she added, "Why would we dry any now when we can hook fresh fish? They taste better." Hearing how blunt she was, Arne grew more dispirited.

One older fisherman stepped in. "And we don't want you and your boats coming over here. One extra mainlander is one too many. We already have a contract to sell all we can land. If you barge in here to fish for yourselves, pretty soon there won't be enough left for us." He was exaggerating, but every head in the crowd around him was nodding in support.

Arne had not expected there would be competition for fish. "Why do you want to keep all your fish for yourselves? I heard there's enough for you and us."

"No," the man shouted. "The fish moved away, and to tell you the truth, we haven't figured out where they went."

Arne didn't believe him. "Listen old man, I'll tell *you* the truth. It's not your fish we want anyway. We said we'd take what you have already dried because there's a churchman back home who is waiting for a load. For us, we're here to trade for walrus. How many tusks have you stashed away? Maybe you still have a pile from last season's hunt."

The Icelander was unmoved. "It'd be a good deal for you, you worn-out pirate, but what's to stop you from coming back for both fish and tusks? Soon we'll all be fighting each other again, worse than the old days."

Arne stared at the wharf. "Stok, how did this entire scheme go so sour so fast?" In disgust, he ordered the sample gear brought back aboard. "The goats, too. They stink, but we're not leaving any part of our load. We've been two weeks at sea now, so it won't hurt to head for Greenland. Dealing with the settlers there can't be worse than here. We're leaving right now."

* * *

On Olaf's signal, the crew sat as one on the open deck. Nobody moved until Arne climbed aboard.

"Get your arses up and grab your oars," Arne commanded. "There's nothing for us here, and we're sailing for Greenland. Now."

Olaf stood slowly, then walked stridently between the rowers. When he reached the bow, he turned and told the others to leave their oars where they lay. All eyes trained on him, then on Arne, then back to Olaf.

"We are not leaving Iceland without some guarantee, Arne. You told us you had a plan and we would all be rewarded. If we end up on another island with no walrus and no fish, where's the reward for any of us?"

Arne gave an anxious flick of his hand. "Think of the rewards you already have, Olaf. Every Viking raider you ever knew has retired to a grubby dirt farm or turned into a smelly fisherman with holes in his boots. We've done better than all the crews in the Norse lands, and I am the reason you have a share of what's stashed on our island at home. If you want to give it all up get off my boat, and stay here. You can take the goats."

Olaf checked with the others. They all dropped their gaze.

"So, pick up your oars, all of you. Your shares of what's in the stash back home are still waiting for you, and there's a good chance of more from what we get in Greenland."

Olaf shuffled to his bench and grabbed his oar.

Chapter 6

Aideen's new dock, Dingle

After the trading depot had been open for a week, a proud Aideen announced there'd be a grand christening. Her two carpenters and the mill workers showed up, happy to enjoy the result of their efforts. A dozen fishermen walked in right behind them, freshly scrubbed after their day on the water. Still in their boots, the men all found sawhorses and barrels to sit on and chatted idly in the empty storeroom. Two farmers from the neighboring pastures appeared, carrying with them an earthy scent. Moments later Lil entered carrying a wicker tray with four marzipan loaves fresh from the manor's bake oven. "Smells like low tide and peat moss in here," she said with a smile. "These little lovelies should help. Can any of you smell these loaves? Two are drizzled all over with our vicar's personal stash of honey, the others are bathed in his favorite rosehip jelly." The crowd cheered as one.

A louder cheer went up when Aideen's cousin Bessie arrived. "Aideen, 'tis a lovely space, but a little chilly. You'll be needing a rousing *céilí* to warm it up, not to mention some of my best brew to bring your new business luck." After she passed around cups of ale she dug into her wimple where her bone whistle had been securely tucked. "I planned to bring my hand drum too, but I couldn't fit it in up there."

The new floorboards vibrated as the crowd step-danced and sang along with the familiar melodies Bessie picked out on her whistle.

"'Tis a perfect celebration," Aideen declared. "If this place is sturdy enough for a step dance, then sure these timbers will handle all the fish you men will bring me. And if my luck continues, it'll handle the special items a certain Basque whaler suggested he might be bringing in too."

After more tunes and a lot more ale, Aideen let the fishermen in on her plans. "I'll be needing you to go out for codfish once a month. How much will we profit from such a tasty trip to sea?" she asked the crowd.

"Not much."

The unexpected verdict came from Sean, a usually quiet man sitting on a sawhorse next to his father at the back of the room. Sean was Dingle's leading fisherman.

Unfolding his arms and edging his cap back to his receding hairline, the fisherman explained, "Aideen, we're out there whenever the weather's fit, but we're not seeing codfish no more. We want to help, but I fear we'd have to go clear across to the Faroe Islands to hook a cod. If it's cod you send us out for we'll be coming back to this grand wharf of yours empty. Where's the profit for you or us? It'd be easier to bring in *scadán* and maybe a sack or two full of cockles."

Sean's response was a blow, but it had made the point the others in the room knew well — the temperature of the sea had risen, and it was being blamed for the disappearing codfish. "Some days we might catch a lone straggler in our nets, but they are rare."

"I hear what you're saying," Aideen admitted, "so perhaps I'll not ask you to do the *catching*. Those Norsemen in the Faroe Islands don't just tend to their flocks of sheep. They are expert fishermen. I've heard they regularly pass offshore of here. What if you sailed out to rendezvous with them once a month, say on the full moon? You'd be able to transfer a boatload from them and bring it right in to me?"

Sean shuffled back to the edge of his perch. "And we hear what *you're* saying, Aideen, but it wasn't so long ago those Norse thieves looted and plundered County Kerry. You know we'll not be trusting them now?"

The local community harbored a quiet hatred for Norsemen, going back to a brutal raid on Dingle's nunnery three generations back. After such a long time, Aideen hoped the old quarrels had been forgotten. "Their days of pillaging are behind them. They've learned trading comes with fewer risks than their savage ways. Anyway, you and me, we're here for fish not friends. And as I hear it, they have loads more cod in the Faroe Islands than they have hungry mouths to eat them. So what if you were to meet them a few leagues out to sea and —."

Sean stood, interrupting Aideen's carefully prepared argument. "Most of us would be uncomfortable going too far offshore, but that's not the point. You see, first you tell us to give up a day of netting *scadán*, our main income. Instead, we're to go farther out and carry back the Norsemen's cod. We'd have to pay them I'm sure. And when we land there's the vicar's tax on imports. Where's the profit, Aideen?"

"Who's to say you're importing it? You rendezvous under the bright light of the full moon, transfer from the Norsemen to your own boats over-the-side, and get paid for the load as soon as you land it here."

Her quick rebuttal caused a few men to scratch their heads. One asked, "What you're saying is we head out of the harbor as usual on the full moon and we return with a load of the Norsemen's codfish. Who's to say we didn't hook it ourselves?"

Aideen relaxed when the boat owners began arguing over who owned the most seaworthy boat and how much they could each carry back. She signaled to Bessie to pour more ale.

Sean took a long swig and stood up again. "On a single trip for you, we could earn the same as we make in a whole month at the *scadán*, but Vicar Maurice sees us at mass.

What'll he say if word spreads his faithful parishioners are now bringing in more cod than they ever did before?"

"Don't worry," Aideen assured him. "He won't have any idea because you're selling to me and I'm not telling."

"That's as may be, but after you settle up with us, he'll read in the parish record where we're all making a larger tithe each month. He'll be wondering where we found the extra income."

"If the vicar asks, tell him it's this changing weather. Doesn't he think he's an expert on all things in the sky, too? Or perhaps you got lucky and found new fishing grounds. What would he know? You're the fishermen. It's your business where you're catching it. It's mine when I'm buying it. And it's never any affair of Vicar Maurice." The men nodded to each other.

"The choice for you is this. You make money selling to me, and your tithe goes to the stewards of our local congregation, or we admit you're filling your boat by importing fish, and Vicar Maurice collects his unfair tax."

* * *

The men all stood, rubbing the kinks from their backsides and getting ready to leave. Aideen knew the fishermen would need to be in their boats before dawn the next morning, so she called out to Bessie.

"Cousin, would you have a last round for these lads? I have another bit of business for them to consider before we finish up." A hubbub arose as the crowd shuffled back to their seats.

"On the dark nights with no moon, you might choose to be bringing back *scadán* or would you take a chance on bringing in the drool?"

"What's drool, Aideen? Is it one last riddle for us?" Sean was sounding impatient.

"Dr-r-rool." She repeated the word with her tongue rolling the Rs. "Dr-r-rool is the code name we're giving the wine you'll be smuggling in." The men and the room fell silent.

"What would we know about *wine*?" Sean asked, more as a verdict than a question. "There may be lots of things growing out on the sea, but I can't say I've seen grape vines."

Aideen understood the man's anxiety. Breaking the vicar's law by importing fish was one thing, but doing it for wine, something none of them drank and only ever tasted from the communion chalice, might be a step too far for them.

"Sometimes you see the foreign whalers in their tall windjammers out here off the coast of County Kerry. What do you think they have in those oak casks lashed on deck? Holy water? No, it's not even ordinary water. When they're at sea so long, water goes all punky, so they carry wine. If you meet a sea captain out there who agrees to trade a keg or two, I plan to stash it up here, in these rafters over our heads." The mention of a hidden stash of alcohol, even unfamiliar wine, grabbed the men's attention.

"These two carpenters have inserted louvers up there to let the misty breezes flow through. It has to be *cool dr-r-rool*, you see." A titter rippled around the room.

Poppie, Sean's father, lifted his cap from his shiny, bald head and held it out. "If I was still fishing alongside these young Kerrymen, I might enjoy rafting up with a foreigner and transferring a cask or two of wine. But how are the lads going to get those wine barrels in here to your cool storage without the vicar knowing? Hide them under their caps, will they?"

Aideen had an answer ready. "You're right, Poppie. Our new vicar has informants who'll squawk louder than greedy gulls if they suspect what you're doing, so when you meet up with the foreigners you'll have your own smaller casks waiting on deck. My plan is for you to fill one small cask each as a gift for the needs of the parish, you know, for the communion service. It'll be your tithe."

Sean was becoming lost in her schemes. "How does all this amount to a profit for us then?"

"There'll be other casks," Aideen explained. "As much as you bring in will be poured from them into extra-long wineskins. You slip them down your trouser legs and into your boots. I'm having leather sleeves made from the hides of special heather-grazed sheep brought down from Galway."

For years, old Poppie fancied himself Dingle's official rhymer. His rhymes were earthy and almost always bawdy, but whatever the occasion, friends and neighbors would feed him the drink and egg him on. This evening, he already had a belly full of Bessie's ale when Aideen's plans gave him inspiration for a new verse.

"Those shepherds from way up Galway talk funny. If your sheep skins have an accent as funny as those Galwegian shepherds do," Poppie slurred, "Vicar Maurice will surely know we're up to no good?"

Poppie's comment made no sense, but several fishermen chuckled as he poked fun at the accent the farmers from the Galway area have. It was all the encouragement Poppie needed. He took two bandy-legged steps out onto the floor, walking as if long wineskins hung down inside the legs of his britches. After just two more steps he had his rhyme ready.

'As I was walking in from Beenbawn Bay, a darling lass stopped me to say,'
'Kerryman, have you a great thing in your trews to make you walk that way?'
'Well, sure I do, but I'm sad to say, 'tis not the big pintle you think it may.'
'This big sack I've hanging in here was made from the sheep of Galway Bay.'

Sniggers and embarrassed coughing filled the room until one fisherman shouted, "A wineskin would be the only thing Ol' Poppie has hanging in his trews." The crowd hunched over sniggering.

Aideen joined in the merriment. "What's in your britches is nobody's business but your wife's, Poppie, unless it is a wineskin you're having in there. Then it'd be my business."

Poppie waddled around the room holding out his empty cup for somebody to fill it. "So it'll be bootleg wine for you," he announced, pleased to be coining a fun new phrase. Several others joined in. "Bootleg wine for you, bootleg wine for us all," they sang, and looped wide-legged around the room.

"Well, I can't spin a rhyme as good as Poppie and I can't sing a note at all," Bessie said, "but I can whistle up a jig for all of *ye*." She played until she had to stop from laughing too hard at the fishermen dancing pretending their boot legs held bags of wine.

Seeing the mood of the crowd was now coming around in support of Aideen, Sean offered a suggestion. "Perhaps you'd be alright if my old Poppie joined you here? He's well past time coming in the boat with me, but he still enjoys being in the fish, and I would say he favors your plans."

Aideen knew the slow-moving man was more talk than work, but it was Sean's way to keep an eye on her. "I'll be happy to have the company," she lied. "At least he might bring a wee slice of fun around the place."

On their way out, Sean tipped his hat at Aideen. "Thanks for the fun evening. We'll do what we can, if just to outsmart our prune-faced vicar."

* * *

The county's newest fish trader and perchance, Ireland's first ever wine dealer, latched the door of her garret and lit a fire in the brazier to ward off the dampness. Stretching out on the narrow straw mat, her head sank into the feather bolster. She expected to fall fast asleep, but as tired as she was, sleep wouldn't come. Questions without end did.

What if the boats don't bring me any fish? What if they do but it's not enough to cover my costs? What if they bring me too many? How many is too many? Will all this drive me to an early grave?

She turned her mind to the maximum amount of fish she could pack into her new storehouse. *I can't imagine how much I'll need to make a good business and how soon I'll need it? Truth be told, I don't have a backup plan in case I never get deliveries at all.*

She was reminded of the wine the Basque whaling captain said he would deliver. *At least there's something, I suppose. Although I don't care for the drink much.*

Chapter 7

Off the coast of Greenland
When they readied the longboat to leave Iceland, Arne watched for new signs of bad weather. Ever since battling the cyclone after they left Aarhus, he regretted not paying more attention to those signs. With an eye on the horizon, he watched Iceland's snow-capped volcanoes disappearing behind them. A verse from an old saga came back to him. 'When those peaks you no longer see, 'tis two more days before the mountains of south of Greenland on the horizon will be'. He was counting on the skies staying clear long enough to sight the peaks of Greenland.

The saga proved correct. In under three days he spotted land and could see where the big island got its name — it was, in fact, green. Arne turned the longboat south and followed the heavily treed shores for another day. When they had sailed past what he decided must be Land's End, he swung to the right and headed up the coast. His objective was the fjord of Eirik the Red, the most famous of all Norse scoundrels, the one who established his legendary Eastern Settlement two centuries before.

The longboat continued north past several wide inlets while fighting a strong current. It was one rocky point after another until Stok noticed a simple indication of human presence. "Look there," he shouted. A scarlet wool rag tied atop a pole planted on a grassy island. It was spiked between the blackened cobbles of a firepit. "There's the marker you seek Arne. It's what they did, those first Vikings. They found their way across the unknown sea and once they arrived they marked the entrance to their settlement with a scrap of red sail cloth."

Arne turned to Stok. "So, they found their way across the ocean. What they did is no better than what I've just done."

Arne maneuvered around the island and followed the deep channel into the interior of the storied land. Before the day's sun was at its peak, the fjord opened up to gentle grassy slopes dotted with hardy scrub trees and strewn with yellow flowers. A ewe, with an already thickening coat, nibbled at the fresh grasses and ignored its two spirited lambs butting heads. Again, Arne was taken aback by how green it was. "This land is so green!"

A cluster of robust homesteads was sprinkled across the open pasture on one slope, each cottage with a plume of smoke spiraling into the cloudless sky. The structures bore the appearance of hard-won prosperity, and the familiar smells of boiled dinner drifted down to the shore.

The crewmen rested their oars and gazed at the settlement. Olaf pointed to the humble stone and timber cottages with sod roofs.

"Want to hear why they build on only one side of the fjord and not over there on the other bank?" he prodded.

"First you are an expert on boat building, now you're the man to give us a lesson on cottage construction."

Olaf ignored him. "They place their homes and gardens on the north slope of the bay so they face south. It gets added sunshine on their homes and vegetable gardens. See, the flower boxes under their windows are sprouting bunches of leafy herbs. But over there to the south, there are still mounds of snow hiding in the hollows and under the brush. There's none around the cottages."

Arne turned to scan the shoreline, hoping to spot a busy wharf, stockpiles of walrus tusks and storehouses full of dried fish. There was one pier — long and broad with space to tie up half a dozen vessels, but only one lone karve was tied up alongside. "The small boat there has the number four painted on its side. I expected to see more of those type vessels used for walrus hunting, so the other three boats must still be up at the ice, killing walrus and gathering tusks." He checked the shoreline again but couldn't locate anything like a tusk warehouse. "Stok, take the boat right into the wharf. They must at least have a root cellar nearby with a store of dried fish."

* * *

When the longboat tied up at the pier, the reaction from the locals was the same as at the last two ports. Residents gathered around and speculated why this unscheduled transport vessel might have arrived. One man, who spoke as if he was accustomed to taking charge, walked to the front of the crowd and demanded, "You're not our regular transport vessel. What's brought you here?"

"We're here to trade a load of tools and equipment for as many of your walrus tusks as this longboat can handle." Arne turned to his crew. "Unload what we brought, including the goats." Turning back to the man he asked, "What's you highest bid?"

"We could always use the livestock and the tools, but we have no walrus tusks to trade for them," the man explained.

Arne's shoulders sagged and he spit over the rail. "This whole trip is making us out to be village idiots. Well, you must have plenty of dried fish in storage."

The man scratched his head. "We have some dried fish left over, but nobody has any fresh fish yet."

Arne jumped down to the dock and marched right up to the man. "You seem to be in charge here. What's your name?"

"Ragnarr Tursk. I'm the Shore Captain for the Eastern Settlement."

Stok was standing at the rail and heckled, "Why does the shore need a captain?"

"The shore doesn't," Tursk snapped back, "but I'm too old for walrus hunting and the residents put me in charge of getting the boats ready. I manage the supplies needed beforehand and handle the market after."

Arne pointed a finger at the man. "If you're the manager, then explain to me why you have no ivory to trade for these valuable tools we brought all the way from Denmark?"

Tursk was unsympathetic. "As for tusks, we didn't bring any home this spring. We were damn lucky to bring most of our men and one boat back after the other was crushed in the ice. And our three fishing boats, they have already left for Vinland. They'll return with enough fresh fish to fill our storehouse, but not enough left over for you."

A scrawny mutt cocked its leg on Arne's boot. He kicked at it, cursed after his foot missed its mark, then marched off.

* * *

Indifferent to the bargaining going on around him, Niels stood up at the prow of the longboat, brushing the wrinkles and stains out of his robe. He put one arm around the carved serpent's head and hauled himself up to his full height. "I never planned to set a foot out of Aarhus, but now I'm staring at the exact community described in our most famous saga." He spotted a tall, broad-shouldered lad in a baggy wool tunic, wearing the muddy boots of a farmer. "You there," he shouted. "I've read there's an actual chapel in this village built by Leif, Eirik the Red's son, for his wife? Which way to visit it?"

"Up there. We haven't had a real priest from Norway here for an age, and nobody uses the place as a chapel, but there it is. I brew all my ale for the settlement in it now."

Pausing long enough to collect his satchel, Niels scurried to the longboat's rail. "This voyage continues to bear wonderous fruit. As a new priest, who could have predicted these blessings I am receiving? What's your name, lad?"

"I'm Benji Leifsson," he shouted and pointed to a rough path leading up across the meadow. "Follow the path over ...."

The farm boy stopped when he noticed the goats in the tiny enclosure. Pointing at the pen, he said, "You're not the guild's regular provisioning vessel, but I see you've got two fine young animals there. I'd be willing to save out a puncheon of my ale in trade for those beasts."

Stok was listening to the exchange and scoffed. "Run along, boy. We're here for a much bigger prize. Take our priest up to your little brewhouse if he wants, but these animals aren't for you. One would be worth a whole boat load of your brew."

* * *

Benji led the priest up the well-beaten path to the abandoned sanctuary. Sprays of spring flowers blooming around the site made it resemble a toy maker's model of a chapel, and Niels was enthralled until Benji threw open the door and dashed in. The scent of malted barley rushed out.

At the threshold, Niels bent his right knee right to the floor. When he rose, his eye caught sight of the brewing paraphernalia and empty mash pots scattered throughout the once-consecrated space. Seeing the clutter and the original hand-carved altar tilted awkwardly against the wall, his heart sank.

In spite of the shambles Niels entered and announced, "I am humbled. This was the first actual sanctuary built on this side of the sea." He became immersed in the religious significance and kept saying, "The first actual sanctuary. Just imagine!"

Four settlers appeared at the entry and asked if they could receive a blessing. Benji cleared the space and they filed in. Niels laid out his missal and chalice with care, along with the few wafers he had packed, a paten to hold them, and a miniature bell to signal the ritual. He checked the pages of his manual to be sure he correctly followed the guidelines.

"This is a triumph for me. There is no gold encrusted crucifix, nor ivory carvings, but I do believe I may be the first Dane to conduct a service on the far side of the sea." His chest puffed out as he crossed himself, knelt and gazed upwards. "I am truly humbled."

* * *

Standing quietly on the settlement's wharf, Tursk was surveying the growing pile of tools coming off the longboat when Arne ambled back and asked, "So if you manage the supplies for the community, what can you offer us for all this?"

"We have little to offer at the moment. Maybe if you wait until fall? The fleet might make one additional trip for fresh fish before they go out for walrus. If they do, you'll have dried cod and maybe some walrus tusks. Of course, the weather might turn and it'll be too late for you to go back home this year."

Arne scowled at the prospect of spending a long, dark winter in the tiny settlement. Tursk noticed and was relieved to think he wouldn't be stuck with the longboat crew for the winter. He did want to secure a deal for the ironware being stacked on the wharf however.

What do I offer a Viking who could take what he wants anytime he pleases? he asked himself. He glanced down at his boots, trying to think of a different proposal.

"There is one other idea. What if you met our fishing fleet down in Vinland now? They could fill your longboat with what they've already caught and stay for more. If you hurry them back to us, you'll receive all our dried fish in storage from last year in exchange for the fresh fish you bring in, plus both goats and all the supplies. Call it a fair trade."

Arne frowned. "If we do go to Vinland to find them, how will we know where your boats are?"

"Head due south. Well, maybe a few points west of south. In three days, you'll see a huge bay. Continue deep into the bay and follow the nearshore around to the east. You'll spot them soon enough. They'll be drifting around in vessels a little longer than yours, but with the same striped, red sailcloth. They stay as long as they want because the weather is warmer down there. If you put the fear of a Viking warrior in them, they'll finish up sooner, plus you could get their extra fish."

For Tursk, the scheme was simple. He left his directions purposely sketchy, and if the Viking crew became lost, they could hardly return to claim their stack of trade goods. On the other hand, if they did come across the fishing fleet and returned from Vinland with a full load, the farm tools would still remain in the settlement.

***

Arne wasn't troubled by the sketchy directions. In his mind, he had enough confidence and navigation skills to locate three fishing vessels, even in an unfamiliar bay, but he regretted trying to make a fair deal. He had one last suggestion.

"Where is the breeding ground of the narwhal? We could hunt a load of their tusks instead. A craftsman somewhere will make chess pieces or church beads out of them."

"And you would be quite wealthy, but the narwhal moves north from here, too," Tursk said it as if he were talking to a child. "They migrate farther away than the walrus. Anyway, a trader from Norway has already collected the few tusks we collected. Both the narwhal and the walrus will be back in six months, but the codfish are waiting now. Which do you want?"

"We don't have much choice, do we? But maybe you could offer a guarantee. Give us your best man to lead us to the boats you say are down there."

"We have no extra men. This is a community of hard workers, and we all have something to do. If not, we'd have perished long ago." Then he remembered the lad who brews their ale. "There is Benji. Benji Leifsson would be the right one for you. The lad makes good amber ale, but there's too much to drink in the village. I'll go tell his grandparents their boy has a job guiding your longboat to Vinland."

"A guide would be good, as long as he's willing to pull an oar."

"Do we have a deal for all these supplies then?" Tursk held his blank expression.

"We have a deal." Arne snarled. "Go fetch the lad." Tursk turned away, ensuring he was masking his smile and set off to find Benji's grandparents.

"Stok," Arne barked. "Move the rest of this load down onto the dock and ask around for enough lye to scrub the goat shit that's ground into the deck. And fetch our passenger again. Tell him we're leaving. The best news for him is that he might see those *skræling* he was told to find. Oh, and we'll be carrying another passenger. The Shore Captain is getting us a guide."

Stok clomped up the path to the tiny wooden chapel and barged in yelling, "Pack up your gewgaws, priest." Niels was hearing private confessions, but the rude intrusion convinced the locals to scatter. Stok held the door open for them as they scurried out. Niels packed up his kit and followed.

* * *

Back at the wharf, Stok boosted Niels over the longboat's railing and onto the deck. When he climbed awkwardly over the rail himself, he noticed the boy from the brewery standing with his family. After the lad hugged his grandparents he vaulted over the boat's rail. Stok was impressed with how agile the lad was and clapped his hands.

Benji saluted back. "Easy for me, but harder for you, eh Pegleg? I'm the boat's new fish finder." He tossed his sack of belongings to Stok, upending him.

Stok regained his balance and hurled the rawhide bag back at him. "I'm not the boat's boy, Greenlander. If you are to be our new fisherman, scrape the goat shit off your boots and stow your gear under the bench over there."

Benji checked behind him to see who was watching and saw his grandparents standing arm-in-arm on the pier. He called down to them. "I guess it wasn't the best way to begin my first trip to sea, huh, Grandmama?"

"We're not worried for you Benjamin. You do need to learn a few lessons as you make your own way and Tursk says you won't be gone too long. Safe travels, dear boy."

Arne readied his steerboard and barked, "You, new boy in front, get your oar wet. Show us you're not another seasick passenger." Benji leaned forward and pulled back, but the oar rotated in his grip and the long blade sliced through the wave, thrusting him back into the lap of the rower behind. Arne snorted.

"I'll help you," offered the oarsman beside him. "My name's Dag. Rotate the blade a quarter turn and grip it tighter, you'll do alright." Benji turned to nod his thanks.

The stern of the longboat swung to the right and the vessel pivoted around the outside corner of the pier. The bow became caught in the swift stream and was carried all the way down to the mouth of the fjord, where they met freshening winds and the cool northern current.

"Captain," Benji shouted. "It's late spring, but there might still be ice down here. Floating hunks are carried along with the southern current. We'll need those currents to carry us to Vinland, but ice could crush this hull. You better lower the sail and double the watch. We have to keep watch all day and at night too."

"Fine," Arne muttered. "Pull in your oar and stand up there at the bow. Keep your eyes open and don't talk to me unless there's a problem."

***

For three days the skies remained somber and the winds blew steadily. Arne followed Tursk's directions, *Head due south ... maybe a point or two west of south,* while Benji maintained his constant vigil. At the one spot where they did encounter a flotilla of icebergs, Arne called for the crew to reduce speed and swung wide around them.

Late on the third afternoon, a range of low peaks appeared in the distance. Benji first heard the cackling and then spotted flocks of puffins taking advantage of the updrafts before plunging out of sight into the sea. He forgot his orders to watch for ice as hundreds of small, colorful birds zipped back and forth. "Those are puffins, the birds I was waiting to see. Look at them all."

Olaf joined him at the bow. "Whatever you call them, they're awful stumpy. Those thick beaks make them top-heavy, but they must need them to hold onto the codfish. Right lad?"

"Nah. You're wrong. They don't eat our fish. But they eat the same little fish that the codfish eat. If those birds are here, our fish will be right below them. And there's another sign." Benji pointed to a harem of seals lounging on a rocky outcrop. "They don't miss many meals. If those seals are hanging around, there's got to be lots of codfish to keep them fed."

"Right on time too. I'm hungry," Arne announced. "We'll drift closer and if you're right, it'll be fish for dinner. A reward for the man who hooks the first one."

The vessel drifted to a stop and each rower took a spool of weighted line out of his kitbag, removed the cork from the hook and tossed the line over the side. Benji made a show of pretending to kiss his hook before he flung it over the rail. He felt an immediate tug. With a whoop he hauled back on the line, hand over hand, until the weight on the hook fell away. He toppled backward onto the deck. His hook came flying up after him, a ragged fish head still skewered there.

"You win Greenlander," scoffed Stok. "It's not much, but Arne did promise to reward the first person to hook a fish. You can lie there on your back and relax; we'll do the fishing from now on."

"And Benji's reward is that he can clean all we bring in," Arne announced to the crew.

Benji regained his footing and glanced over the rail where a seal circled on the surface. The rest of the huge fish was hanging from one side of its jaw.

Benji coiled up his line and squatted amidships. His disappointment with his 'reward' was embarrassing until Dag, the other young oarsman, squatted with him. The two cleaned and split each fish that came over the rail.

Each time a fresh fish came aboard, the whole crew cheered — except for Niels. He lost his appetite as soon as he saw Benji and Dag covered in fish guts. Hanging his head over the side, he groaned, "I once said I enjoyed fish, but this is different altogether."

Chapter 8

Fishermen's Quay, Dingle
Aideen checked the line of boats for the stout vessel she expected would offer a pleasant overnight cruise upriver with Paddy. She saw only small currachs tied at the fishermen's quay and wondered where else Paddy might keep his boat. When the easy-going sailor ambled over, he smelled sour, no better than yesterday's ale. He casually pointed to his boat, an open currach bobbing listlessly beside the fishermen's quay. "Hop down in, dearie."

"Paddy, have you been into the ale all night then? And what might be more of a problem — is it your plan for us to go all the way to Limerick in this little craft?"

"No I haven't, and yes, it is. I made this little beauty myself. Seventeen of the best ox hides went onto her frame, all stretched over a stout wicker frame. She's done me proud over many a league."

"It may be fine for you to travel up and down the coast in a boat made of skins and wooden hoops, but no sea nymph nor a selkie am I. I don't even swim."

"Aideen, I made this currach broader than most, so it's fit for carrying goods plus a passenger or two. Once we hoist the sail and we're gliding along with the tide, you'll see how comfortable it feels."

He tossed her sacks up in the bow. They each landed with a soft thud and the boat took on an acute forward tilt. Her fears grew until Paddy settled himself in the stern and the boat returned to level.

"Settle yourself in the gap between those two coils of hemp rope, dearie. Hold onto the side rails if you want." She clenched her stomach muscles, stepped down onto the coil of rope and gripped a gunwale. The small boat eased out into the harbor and Aideen used her free hand to draw her wrap tight around her shoulders.

Outside the harbor, a late evening breeze made the sail go fat and the currach picked up speed. The cool spray splashed up to her knuckles, but cold or not, she was determined not to release her grip. The rushing water underneath her concerned her more. It rippled against the taut animal hides.

"The old seafarers say the skins have to move like as if the oxen are still breathing." Aideen wondered why she ever left the security of dry land and Paddy's commentary was little comfort.

At the mouth of the River Shannon, the breaking dawn brightened the sky, but not Aideen's mood. Watching the ripples in the currach's skin, the only thing between her and the water, she held her grip on the gunwales for the rest of the trip upstream to Limerick.

* * *

"We'll be alongside the wharf soon," Paddy announced. "Before we get there I should mention my expenses. You see, if I don't secure a return load, this is going to cost me more than I will earn by trading your sacks of beans."

"You needn't be worrying, Paddy, my man. If this trip goes as I plan, you'll be making many more visits — and profitable ones, too. On your next trip for me, I'm hoping you'll have cod fish from the Norse fishermen here and even some kegs of wine you'll say you bought for me at a Limerick marketplace."

"What do you mean I'll *say* I bought some wine? How many casks will I buy?"

"None," she answered. "You won't have any at all. You see, when Sean and his fleet collect a few casks of Spanish wine out at sea and smuggle it into me, I'll be hiding it up in the rafters. If anybody finds out, I'll say you brought it down to me from Limerick."

Paddy stared at her, puzzlement obvious in his eyes. "Aideen, I want you to succeed, but pretending I have wine won't work for me. I have to pay for my meals at Bessie's. And I don't understand how you'll earn any money at all for your effort."

Aideen took a breath. "Remember, my friend, it's not how much we'll be making for ourselves. What is important here is getting the best food and drink for our people in Dingle, plus making sure none of us pays the vicar's taxes."

Paddy couldn't mask his disappointment and changed the subject. "You told me you wanted to connect with some Norse fishermen, well Limerick may be the place to find them. The trouble is, after the Normans invaded, they forced the old Vikings into the shadows. These days Limerick is divided into separate villages. You'll have to find somebody who knows where the Norsemen are and if they still have contacts with the Faroe Islands. Limerick has lots of back alleys and a few rough spots, mind. You'll not want to get yourself lost."

The currach glided smoothly into an empty finger pier where it was dwarfed by the masts of much taller vessels, each clattering with the thwack of cordage against their masts. Behind those larger boats, the high turrets of the massive castle being constructed right at the river's edge overshadowed everything.

Eager to extract herself from the uncomfortable perch between the coils of hemp, Aideen climbed up to the wharf, stamped her feet to ease the cramps in her buttocks and gazed at the castle's weighty walls.

"Those walls are thicker than a man is tall," Paddy explained. "The Norman overlords say they are for our protection, but we know they're meant to show how the English are in charge here. You'll be welcome at the bazaar inside the walls, but you'll have to speak the king's English and the shopkeepers are always hard arses to bargain with," he warned. "And as I told you before, if you go looking for the Norseman, be careful you don't get turned about. If you plan on coming home with me, I'll be right here at the quayside, ready to leave when the river rises as high as the top of the bridge pillars there."

"I'll not get lost among my own people, Paddy." She turned her back on the waterfront and headed into a labyrinth of narrow alleys.

* * *

Aideen noticed how each passageway leading away was alive with boisterous street urchins and barrow pushers. After several blind corners and a dead-end alley, she was indeed lost. Hearing workmen finishing a parapet on a castle tower, she stepped backward to see the top and bumped into the ashen stones of another wall. The arched gate and each stone had the scribed signature of the Irish mason who fitted it there. The Celtic cross of St. Patrick was chiseled into the keystone.

This has to be a nunnery, Aideen decided. *Bleak walls such as these are surely built to protect our good sisters inside, regardless of who invades this land.* Aideen made the sign of the cross, rapped twice on the thick entrance doors and waited until a sister slid open the peephole in the sturdy timber.

"*Dia dhuit*. God be with you. I'm Sister Enda."

Aideen was comforted to hear the traditional Irish greeting and to see the nun's nervous smile through the peephole. "*Dia is Muire dhuit*. May God and Mary be with you," she replied.

Dressed in black from head to toe, except for her white coif and neckerchief, the nun swung one of the doors ajar. Her wool tunic swayed noiselessly as she led Aideen through the cloister into the vestry. The comforting fragrances of the candle wax and incense filled the space. Aideen grinned at how nuns always walked so quietly. They reached a bench, and Sister Enda suggested they sit so she might ask how she could be of assistance.

"Sister, may I first commend you on taking the name of St. Enda. I'm told he was a credit to his kin in the Aran Islands."

"He was my inspiration from an early age. I can only hope to follow his righteous pathways."

"But tell me, Sister, Vikings oft visited those remote islands. Are there any of their Norse kin left now? I'm trying to find a codfish vendor and thinking nobody but a Norseman might know who has some on offer."

"This part of Limerick is now called Englishtown, my dear. All the tradesmen in the cottages here have been brought over from England to cut the stonework on the castle. The Norsemen and your Irish kin, I mean *our* kin, have been herded across the river. You'll have to cross over the River Abbey to Irishtown. Take the Baal's Bridge. It's the narrow, humped bridge up the lane there. No farther than a rat can scamper, it is."

"You say the Normans forced the Irish to move out of their own town, Sister? Did the old Vikings go with them?"

"Ah, you're in luck there. After our chieftains drove off the worst of those awful raiders, some of the peaceable Norse tradesmen remained. There are sure to be a few of them around, the ones who settled down with locals and had families, I mean. They'll be living over in Irishtown. In fact, it was the Vikings who built the bridge over the River Abbey."

"Well, I do speak a little of their language. If I give them a try, could I trust them to deal with me?"

Sister Enda explained the respectful peace they now had with their old foes. "I suspect you'll be alright. I admit there was no love lost between us and them for a good while, but relations are fine now. Both sides agreed it's better to work as partners to deal with these Norman invaders. But tell me, dear, why would you be seeking cod? It's *scadán* the Norman lords want. I can't tell you why. Give me a salmon anytime, especially during Lent."

"Sad to say, Sister, but we in Dingle have precious few of the salmon you have here on your doorstep. We think of codfish as the next best thing. The difficulty for us is finding one."

"Ah, forgive me, dear," pleaded the sister. "I do forget there is a real world outside our gates."

Aideen gave a slight wave of her hand. "Don't worry. In truth, Sister, some things on the outside you needn't know. But salt fish has become a quest for me now."

***

At that moment, a statuesque, pinch-faced figure in a starched habit appeared in the archway. The abbess of the convent swished straight down the corridor and looked Aideen up and down. "I'm kept informed of visitors under our roof," she said courteously. "What is it we can help you with my dear?"

"As I was telling Sister, we in Dingle have none of the salmon you have here in the River Shannon. We think of codfish as the next best thing, but they are rare. My intention, Reverend Mother, is to find a trader here who can acquire codfish for me."

"Trading in the fish is it? You must be aware, my dear, both buying and selling in our town require you to pay the tithe." She explained the rule sympathetically, but what she said next was more blunt. "And since we already have a call on the salmon caught in the weirs below Limerick, it would be unusual for us to accept cod as payment."

Aideen had neglected to include any payment to the church in her plans. She was now trapped by her own words. And worse, any tithe given to the nunnery would become a matter of record. The clerics in the parish here would surely tell Vicar Maurice of it. She needed to think quickly.

"It wouldn't be fish I'd be offering you." She clasped her hands together hoping what she planned to say next was proper. "I might have a different contribution. If it pleases you, our plan is to land wines at my new wharf in Dingle, in part to address the needs of you and your family here. My concern is that we will have to pay our Vicar Maurice's levy on each keg we bring in. Perhaps you could explain to your bishop here in Limerick and have him inform the Vicar General in Dublin about my plans to assist you with God's work. Once he is made aware, the Vicar General will surely waive Vicar Maurice's import tax, and I will be able to support your cloister with generous quantities."

The abbess grinned, showing two full rows of teeth. "Ah. An offer of wine sounds a sight better than any fresh salmon. I am pleased to hear of your interest in assisting our family my dear, and I can assure you the man's tax will be annulled."

"I am eternally grateful, Reverend Mother," Aideen said and smiled back, thinking; *Now let's see the new vicar dig deep enough into my purse to come out with his import tax.*

Content to think she had at least countered one of her problems, she stood and said, "Alas, since codfish has become my quest, perhaps my next stop should be over the bridge in Irishtown. With my thanks to you both, I'll be on my way there, and then I have a boat ride for home to connect with this evening."

The three recited a traditional short travelers' blessing. *"Slán leat,* safe travels." Aideen politely begged her leave.

Sister Enda caught Aideen's hand as she stepped to the big gates and whispered, "If you don't make it to your boat ride and you're needing a place to stay, come back here, no matter how late." Aideen thanked the sister but assured her she would be fine.

*   *   *

Aideen headed for the Baal's Bridge. Within moments, she was again confused by all the narrow alleys. Hopelessly turned around, she stood stock-still when she sensed a moldering whiff of green algae. She heard a stream babbling over a disused fish weir. After a few steps toward the sound, she was within sight of the humped bridge crossing over the slow-running River Abbey. Feeling like she was going home, she crossed into Irishtown.

A dusty black form, scarcely more than a fur-covered skeleton, wrapped itself around Aideen's ankles. It purred louder than any cat Aideen had ever heard. "For a scrawny animal your purr is louder than a lion's growl. Maybe you know where to get a lion-sized feed of fish."

A little urchin bent down to corral her cat, and Aideen asked, "Where's your village market dearie? Is it where I'll find the local traders who might be into the fish?" She was directed to a loose jumble of carts and stalls. There she came upon another waif sitting proudly amidst seven barrels of salt.

"Here I have the freshest of sea salt," said a tiny girl in a voice trying to imitate the throaty bark of more mature proprietors. When she noticed Aideen's interest, she continued, "Collected by Mama's kinfolk on the Aran Islands, it was. My uncle brought these barrels in scarcely two hours ago. I expect some pig farmer will snatch them up if you don't take them straight away. What will you offer?"

"I can't imagine this all will be hustled away so quickly, but it's from the Aran

Islands, you say? This must be a sign from above telling me to take you up on your offer." The girl's mother was standing in the back, letting her daughter charm prospective customers. Her daughter had succeeded.

"Tell me, little one," inquired Aideen. "If I was to offer you a sack full of the finest Dingle beans, it should be enough to make you give up all this salt. Am I right?" The little girl looked back at her mother who winked her approval.

Aideen responded with a friendly smile. "There'll be a boatman named Paddy down at the quayside nearby the castle. He has a sack of tender field beans waiting for you and if you tell him I sent you, he will come back with a dolly to collect all these. Would such an exchange be to your liking?" Without another word the girl vanished into the crowd. Aideen thanked the girl's mother and asked if there may be peddlers in the market who handled cod fish. The lady shook her head.

"More is the pity," Aideen said. "I'll keep wandering these rows of stalls for a while then."

After an hour of back alleys and unhelpful suggestions, she remembered she was supposed to meet Paddy at the wharf. *He'll be loaded by now and ready to ride back down the river*, she reminded herself. *This trip is my one chance. I've come all this way and I do have a load of salt, but not a fish to put in it.*

Discouraged, Aideen turned back for the docks, found Paddy drumming his fingers on the gunwales of his boat, and hopped aboard.

"I wasn't going to wait much longer, Aideen. My business runs with the river and when it's ready to go out I can't be sitting here, paying passenger or not."

"Well, I was able to solve one problem for us, Paddy, my man. The abbess at the nunnery says she will get the vicar's tax on wine canceled. At least you will have some actual wine deliveries to make to her and not be saying you have wine on board when you don't."

"Good thing Aideen, for I never did understand any part of what you were saying."

"And then there's the salt too, Paddy. I see you have it tucked aboard here, so a few good things are coming our way."

Chapter 9

Notre Dame Bay, Vinland

Niels' body ached from vomiting all night and the day before. By dawn the deck around him reeked of goat droppings and puke. He squatted in the shelter of the forepeak, shivering under a fleece hauled tight around his shoulders. Frequent gusts of damp sea air snuck underneath, sending chills up his backbone and forcing him to pinch the material around his neck more tightly.

When Arne noticed the morning light breaking through the fog, he shouted. "Hey, priest. You may be a passenger, but you'd be a more comfortable one if you stood up and moved around. While you're at it, climb up onto the serpent's neck and watch out for rocks." Niels did what he was told.

An assortment of islands dotted the surface, each with rocky shorelines rising to low knolls crowned with thick, dark green trees. His attention was taken by a cormorant standing on a rock drying its wings. He leaned over to watch it take flight and spotted dark ledges lurking below the surface.

"Rocks, Arne," he shouted, "and jagged shelves over there, breaking the surface."

Arne made out the black shelves, cursed and barked, "Stok, hold firm on the steerboard and keep us away from any rocks you see. The rest of you check for the ones Stok doesn't see. With all these islands around us, I'm betting there'll likely be shallows and sunken ledges too numerous to avoid."

* * *

As they rounded a headland, clouds of seabirds could be heard, then seen, circling a trio of longboats.

"I'll wager it's the Greenlander fleet I see over there in the cove. We'll drift over to them, but slowly. Dag and Benji, hold us steady with your oars, the rest haul your oars in, lash them under the gunwales and help drop the sail."

The longboat drifted closer, and Arne could see the Greenlanders tossing fish overboard. "It's how they get only the best fish, Captain," Benji explained. "They're hooking one fish after another after another. They only save the fattest ones. The smaller ones are tossed overboard. They also do it when their holds are full because fishing is fun and nobody's in a hurry to go back to Greenland. I've heard they sometimes pull up two fish on a line because a small one bites and a greedier one swims up to swallow the first one already on the hook. It's hard work hauling them both up with two on the line."

Arne shouted over to the three vessels, "I made a deal with your manager back in Greenland. He wants you back at the Eastern Settlement now with all the fish in your holds. We'll fill up with the fresh fish you are getting now."

The fishing captains shouted for their crews to bring up the older fish from deep in their holds. Stok picked up an oar and shouted, "Not yesterday's fish, we want the fresh ones."

"You men have a way of getting what you came for," conceded one captain. He relented and ordered the crews to transfer fresh fish. "We'll give you our top layer and refill our holds later."

Arne was still irritated. "Those fish are huge. Start filling our hold now. We're all returning as soon as our boat is loaded."

* * *

Niels tried to keep himself out of the way, but he too was annoyed by all the wasted fish. He scolded the fishermen as he would bad children. "We have come all this way for your fish because we don't have enough for our parishioners at home, not to mention any to waste." He turned to Arne.

"I'm pleased to hear you share my concern for this God-given bounty, but remember why we are here. There have long been rumors of people living in these forests. I need you to put me ashore to verify those rumors."

Arne shrugged, but one of the Greenlander captains agreed. "We can help you, Priest. I plan to send our dory ashore for water before we leave, so maybe your captain will let you go in with it."

Buoyed by the first show of support for his mission and with renewed self-confidence, Niels wrapped his satchel over his shoulder and stood, patting the wrinkles from his cassock. "Thank you. I'm ready."

"Before you go, priest," the fisherman continued, "you should know the Forest People, the ones we call *skræling,* are afraid of us. We have seen them fishing from canoes in this bay, but they disappear deep into the woods when we go ashore. You won't see any of them, and my men will not be going into the woods to find you when you get lost. Do you still want to go?"

"As a priest, I am obliged to bring blessings to all peoples. Perhaps they will welcome me if I make an offering of these discards. Would you please box a few up for me to carry in for the poor souls? If they are hungry, this will open their hearts to the word of our Lord."

Arne snapped. "Benji will go in with you to keep you from getting lost in the woods. We're leaving here on the high tide, so when this little boat returns, you better be in it."

"I'll help too," shouted Dag. "Let me go with them in the dory."

Arne turned his back. The two men launched the dory and dropped in a pair of empty barrels.

"Hitch up your dress, priest, and step over the rail. Sit up at the bow," instructed Benji.

"And the fish?" asked Niels. "I should have an offering if I connect with them. They must not think of me with any fear."

Arne waved his hand. "Fill up a box for him. The birds don't need any more of it."

* * *

More carefully than an alley cat stepping over puddles after a downpour, Niels hiked up his cassock, gripped his satchel, and climbed down into the bobbing dory. The last step was farther than he expected, and he fell in a heap to the bottom of the boat. After the others dropped in and helped Niels right himself, the heavily loaded boat skimmed through the carpet of fish carcasses, headed for the beach.

Clutching his special medallion and dutifully thinking of Saint Christopher, he explained, "Bishop Olsen's gift reminds us to think of this rocking up and down as if we are cradled in the arms of Notre Dame herself. We ask him to protect us from all the danger in this huge bay of hers."

Benji rolled his shoulders and continued to row against the waves. "I guess the dame from Notre Dame might decide to help you, but tell her not to worry for us."

The bow of the dory nosed into the purple pebbles. Benji and Dag threw the water barrels and the fish box on the shore and set off to locate a brook. Niels hiked up his wool robe with one hand and used the other to grab a gunwale as he stepped out into the surf. "Wet feet be damned," he shouted, with a sense of personal accomplishment. "I have arrived." He wobbled over the loose footing to the high-water line and knelt in a windrow of dried seaweed to give silent thanks.

Out loud, he asked himself, "What are the first words I should say to the woodland dwellers?" A few inspirational phrases flooded into his head. "Perhaps I should recite a prayer?" He bowed his head to recite familiar verses and ask, "Will they understand anything? Perhaps not without having a proper introduction to the scriptures first."

He stood, blinked and noticed a structure of driftwood woven with thin tree roots in a youthful attempt at a raft. It was still dripping.

Speaking as if he were directing a search party, he said, "It must have been pulled up onto the beach minutes before. Somebody is here." His mind raced. Peering into the dark tree line, he announced in a louder voice, "If you are watching, come hither. We bear you no harm."

He tried again, hoping that if he sounded more supportive, whoever was hiding behind the tree line would walk out and greet him. "And be reassured by my pious demeanor."

Nobody appeared. He called out in a more condescending and pastoral tone. "The Greenlanders say you often wait and watch. They tell me you are too shy to make contact, but we have all this bounty of the sea to offer. If you want it, you should show yourselves now."

Trying to appear as unthreatening as he could, he knelt again, on both knees and with his prayer book in an outstretched hand. An eerie stillness filled the space between him and the tree line, and he struggled to mask the apprehensive grin from spreading across his face.

He was half fearful, half hopeful, and his voice quaked as he asked, "Are you in there watching? Do you hear me above the squealing birds and pounding surf? Your forest of spruce and fir is dark and thick, but do not conceal yourself or fear me. I bring you no harm." Niels stood, brushing sand and dried seaweed from the folds of his robe until the sound of splintering twigs sent a shiver up his spine.

* * *

"Hel ... hello? Please, show yourself." The bushes moved. "I am Father Niels from Denmark. I've arrived here on the Norse trader you see out there. I have gifts to offer. There is some fish and here I have the Good Word." He wondered if he should hold his prayer book higher and chant the name of the Lord. They may not recognize how powerful it was, but it would quell his own fears of supernatural horrors lurking beyond the dark line of trees.

Another sound rustled in the shadows. He froze again, sure there was any number of ominous gazes directed at him.

Two figures stepped out of the bush. Dressed in rawhide and their faces decorated with smudged markings, they walked barefoot and wet up past their knees. Strands of their long hair flowed freely in the wind gusts.

Niels took them to be youngsters and extended an outstretched hand. They both took a step back. The older one kept a tight grip on his spear. Behind him, a shorter, younger lad tried his best to appear menacing.

The older boy waved his arms out across the bay. "*Beothuk*," he said.

Niels had no idea to what the boy's word referred, but gave them his best benevolent priest smile. Engaging his diaphragm to calm his nerves, he said, "Peace ... be ... with ... you." He waited a moment for the required response. When only silence came from the boys, he tried a different greeting.

"I have an offering for you." Walking over to the dory, he pointed at the box of small fish and with both arms outstretched, he said, "I think this is what you call *Beothuk*? Please, have it."

The boys spun on their heels and disappeared into the trees.

Standing alone, the puzzled priest waited for a sign. Nobody had ever shied away from him before, particularly when a gift was offered. This failure to communicate embarrassed him. He was sweating and uncomfortable in his heavy clerical garb. *What do I do now? I wish Bishop Olsen was here to make sense of this for me.*

"Priest," Benji called out. He and Dag had been standing in the shallow surf watching to see if Niels was making any progress. "We are ready to leave," they shouted.

The interruption unnerved Niels. He made the sign of the cross, hiked up his robe and scurried over to the waiting dory. A glistening black cormorant landed on the edge of the box of dead fish and gave a pig-like grunt.

* * *

When the dory butted up against the longboat's hull, Niels clambered back aboard without needing a hand up. He unbuttoned his robe, hung it over the rail to dry and crawled into his nook in the longboat's forepeak. "What a wonderful experience," he said to nobody in particular. "I encountered some of our woodland brethren and will report so to Bishop Olsen. And I will be pleased to add how I christened our landing place 'Notre Dame Bay', the spot where I encountered two of the local residents and filled our boat with fish."

His smile disappeared.

"No. As a priest, my duty was to make an effort to introduce them to the Good Word and at that, I failed." He crossed himself. "Now I must say twelve Hail Marys, hoping the Heavens will forgive me."

Chapter 10

Eastern Settlement, Greenland
The Hail Marys were recited and forgotten. All the way back, Niels prattled on about his encounter and the infinite number of fish in Vinland. When the longboat turned into the fjord, Arne shouted at him, "Enough, priest. I admit you were right when it came to all the fish in Vinland, but our journey is not half finished. It's a long way back to Aarhus."

Leading the three fishing boats into the long pier and tying up, Arne was determined to avoid any further confrontation. "Stok, have Benji and his Greenlander friends unload us. They can cart all the fresh fish off to the settlement's drying racks. After, get them to carry back as many bundles of his dried fish as we have room to carry. Fill the hold except for a narrow passage into the right side, up by the bow. Leave a narrow space empty for now."

"What for?" Stok asked.

Arne spied Tursk walking onto the dock. "I'll explain later."

Tursk opened with a greeting fit for a long-lost friend. "Arne, we have to thank you for getting the fleet home sooner than expected and with such a load. You're now welcome to all the stockfish in our warehouse."

Arne was non-committal. "It's the one worthwhile part of what's been a bad deal for us."

"Well, to be honest, it went better than I expected. Why don't you relax? I'll have Benji bring down some of his ale. He has more than enough for your trip home."

Arne surveyed the wharf, wondering if there was any more he could get from the deal, but Stok butted in. "The man is right, Arne. We're ahead on this deal now. Let's stay and drink all Benji's ale, then pack in more for the trip."

"Not yet," Arne muttered. "Not yet."

Olaf had been on the deck, casually watching, and confronted them. "You told us there'd surely be a profit waiting in Greenland, but I haven't seen any tusks and there's nothing else for us when all this fish goes to the church. Now you say 'Not yet' to the ale. Our share of nothing is still nothing. You're wrong again."

Arne turned his back on Olaf and paced back and forth next to the incoming crates of dried fish.

Stok grabbed his brother's elbow. "You're pacing more than an expectant father and you hate children." Arne wrenched his arm loose and kicked over an empty keg. Stok persisted. "Since you've tipped it over, plant your arse there and tell me what's going on in your head."

Arne kept pacing. "Olaf is right. We can't go back to Denmark and give all this fish to the bishop. There'll be no payment for our effort and we've got little ivory to sell. Look at what we've become, Stok. We've become a boatload of errand boys. Where's the reward for anybody?"

"We have the longboat," Stok reminded him. "A fine and *free* longboat."

"We can't eat a longboat," Arne snapped. "And the crew. Did you forget them? There's nothing to pay the crew shares. We gave away the last of the trade goods to this crowd in return for this fish we've promised to give away."

Arne dropped down on the keg and let out a harrumph. "There's the problem, so listen to what I've worked out. We'll follow the same route back, past Iceland, then to the Faroe Islands. But from there we'll turn south to the coast of Ireland. It'll be coming on the heat of summer there, right?"

Stok bent over to rap his knuckles on his wooden leg. "Ireland, Arne? I have bad memories of our last visit there and one less leg. What do you want to go back there for?"

"In summer there's always a haze along the Irish coast. We'll get lost in a fog bank on the way and fake a disaster to generate panic. And guess what? This will be the way we use the dory the yard workers made for us. The bishop put his priest on our boat to be sure we'd return with the fish, but we'll tell him we're sinking, shuffle him into the dory and set him adrift. There'll be other boats out there to rescue him. Once it's done, we'll take this load of fish to some market town in Ireland. The Irish love fish. We'll sell what we have right there. If the priest ever does get home, all he can report is how we sank and he was rescued. He'll think he was the lucky one. His bishop won't learn what happened to his boat or to this load of fish, *our* load of fish."

A smile broke on Stok's face. "So, that empty space in the hold next to the bow ... are you now saying it's part of your plan?"

"I'm saying that space will be the key to the plan when we announce we're sinking."

Arne swiveled on the keg and pointed to the longboat. "We're going to bust a hole in the hull and fill part of the hold. The boat will survive. This is the finest longboat the Aarhus yard could build, remember? The keel is solid oak, and the hull is as strong as Greenland's summers are long. We'll be fine."

* * *

Niels was still feeling buoyant. He marched onto the settlement's dock and straight to Tursk, who was supervising the fresh fish.

"Mr. Tursk, my mission was to collect accurate information on where the fish are for future excursions such as this. Would you share your estimate of the number of leagues we traveled from here to Vinland, the local currents, seasonal weather patterns and other details you think are important? Perhaps you could write it all down?" He slid a sheet of virgin parchment out of his satchel. "And a map. Maps are always helpful."

"I suppose I could, Father." Tursk sketched a crude map showing Vinland at sixty-three degrees north latitude. He made quick notes in the margins of the distance to Vinland and the prevailing summer winds, then said abruptly, "I have to go and finish supervising the settlement's fish landings."

Niels didn't suspect that each of the details was a guess or worse, a deliberate lie. He turned to Stok. "This is wonderful. I'm thinking this voyage to Vinland was not a failure at all, in spite of not being able to baptize any of the woodland residents. In the future, you and Arne return with more vessels to join the quest for fish. We should then be able to feed all of our parishioners." He bowed and set off up the path to the tiny chapel to give thanks and to pray for a safe return.

* * *

When Benji wheeled his barrow with a brace of full kegs down onto the wharf, he could see Arne and Stok huddled in conversation at the far end. Several paces away, Dag was crouched behind a pile of empty crates. Dag signaled to Benji to squat down beside him.

"I've been picking up parts of Arne and Stok's conversation. It sounds like they plan to keep the fish, carry it to Ireland and sell it there."

Benji perked up. "Ireland, you say? When I was little, my grandmama told me there used to be Norse settlements there. When I got older, she also told me the Irish people brew a tasty blonde ale. There's another reason for me to go to Ireland."

"No, listen to me. This is sounding bad," Dag whispered. "Heading to Ireland to try their ale is one thing, but this is serious. They plan to fake a sinking on the way and set your friend the priest adrift."

Benji bolted out from behind the crates and strode directly over to Arne, who was sitting back on the keg.

Arne glared at Benji. "We're busy. And you can't get us any walrus tusks, so I think you're done."

"Arne, what you're planning isn't right?"

"What do you mean farm boy?"

"I'm not telling you where to sell your fish, but I am saying you can't set the priest adrift at sea on his own. The priest is my friend. You can't set him adrift." Benji knew confronting Arne was a great risk, but he was beyond taking back his words. Arne was instantly angry.

"How did you —."

"We overheard what you're planning," Benji blurted out. "You see —."

"Do you think we need your advice on how to do any of this? Go back to your little boy brewery. We have a fish business to run."

Benji tried a different approach. "I want to join your crew. I mean permanently. Dag and I work well as a team, and if he's going to Ireland as part of your new plan, I want to go with him. If you take me, I won't have to tell the priest, will I?"

Arne eyed Benji up and down. "Thanks for the offer, but we don't need a fish finder anymore. If this load goes as we think, maybe we'll return next year and talk to you."

"No. Not next year. If I go to Ireland with you, I won't tell my neighbors here what you two are planning for Niels."

Stok jumped in. "Arne, he must have a lot of guts to stand there and speak to you that way. We'll have to keep him on our side because if he mentions our plan before we leave, Niels might not be smart enough to believe him, but this community will. They'll descend on us, and we could lose this load. He's learned to handle an oar as well as the rest of us, so I say we keep him."

"Right," said Benji. "So before you put Niels in the dory, both of you make certain you hear boats nearby to rescue him, and two men with extra drinks to go with him in the dory."

Arne stood up and sputtered. "Go get your gear and bring the priest back with you. If you breathe a word of this to him, or to anybody here, I'll put you in the dory with the priest and nothing to drink for either one of you."

* * *

All the way back to his grandparents' homestead, Benji wondered how he would tell them he was going back to sea on the longboat. When Benji first told his grandparents he wanted to be part of the longboat's crew on the trip to Vinland, his grandmother warned, 'Those men are ruffians and those boats aren't built with the comforts you have here at home." It was only when Benji argued it was his chance to learn navigation that his grandfather agreed he should go.

Benji walked into their home and found them felting a sack of unspun wool roving.

He started in with the same agreement. "Grandmama, Arne and Stok are heading to Ireland and I'm going with them. Arne navigates by the stars. He watches the birds, sniffs the offshore airstream, and he has a crystal to find the sun, even if it's behind a cloud. When we're at sea there'll be lots of chances to learn how he does it, and more."

His grandmother responded without hesitation. "Benji, you shouldn't be mixed up with them. Those men are pirates. We all miss your father and we remember how he wanted you to become a pilot. Perhaps it was alright for you to go to Vinland, but this trip will be much longer. What places will they take you to in Ireland? Nowhere safe, I'm sure."

"I plan to keep them honest, Grandmama. I made a friend on board, Dag. He and I are going to take care of the priest."

This time Benji's grandfather frowned. "What you are saying is backward. Priests are supposed to take care of us."

"Most often you're right, Grandpapa. And the priest does try, but there is one favor I have to do for him first."

All his grandmother could think of adding was, "Benji, you have a life here on the farm, not to mention chores to do for your grandpapa and me."

Benji's grandfather winked at Benji. "Perhaps it's best. You've grown to an age when there must be more in your life than brewing ale. You have yet to be with a girl, and there are no young girls here to be with."

Benji blushed and kissed them both. Tears etched tracks down his grandmother's cheeks.

"Always be good, my boy."

* * *

Benji scurried up the path to the chapel. He noticed the two young lambs he had been tending and smiled as they pestered the ewe to feed them. "You're not happy now, and they don't know it yet, but those two will be fine when you wean them," he told the ewe. "The time has come for them and me to be on our own."

When Benji opened the door to the chapel and stepped in, Niels was on his knees, repeating a list of the good things that had come to him so far. Seeing Benji, Niels lit up. "My friend, so much has happened to be thankful for. I may yet earn a promotion."

The words barely escaped his lips before his gaze dropped. He looked as if an ax might fall on his head. "Saints forgive me. Yet again, I am guilty of aspiring to rise above my station. Perhaps if I say twelve more Hail Marys ...."

Benji waved his hand. "You will be forgiven, Father. Anyway, Arne says we're leaving right away. And the best news is, I'm going along with you."

Niels beamed at the announcement. "You're coming aboard with us again? Excellent news. It is such a relief having you along for Arne and his men are difficult most times. I hope I don't cause you any bother or conflict with them."

"My plan is to see you back to your home parish and receive the promotion you deserve, so pack up your things and leave the brothers to me."

Niels carefully fitted his portable mass kit into his satchel along with the crude map of Vinland's fishery that Tursk had drawn.

"This map will secure the future of meat-free days for our parishes back home." The two hurried back to the waiting longboat.

* * *

Benji sat at his bench and threaded his oar through the oarport beside him. As the longboat eased into the fjord's swift current and picked up speed, Niels saluted the tiny chapel on the hill, thinking he was the luckiest priest in Christendom.

"Benji, see how the warm breeze is at our backs. It's so much better than the first days at sea. I'm sure our return will be trouble-free and blessed with fair winds."

Benji nodded.

Chapter 11

Off the coast of Ireland

The prevailing easterlies combined with an unseen mid-ocean current. The trip homeward was faster than Arne expected. He pointed to a bald, dirty-white rock, covered with bird droppings that stuck straight out of the water. "See the tall rock pillar right over there? It's called Rockall. I was hoping we would see it. The sagas say it's a signpost for travelers. They say if you see it, you're lost."

He turned to Olaf and snorted. "You didn't know it was out here in the sea, *Olaf the Ugle*, did you? And I'll tell you something else. It means we're not far from Ireland." Arne knew Olaf hated being teased and he usually avoided angering the man, but with a sale for his fish at hand, Arne was feeling emboldened.

"Stok. There's a change for our course. First thing tomorrow, I want to be right on Ireland's herring grounds."

"I'm keen to see Ireland too," Niels piped in. "I should make ready." Arne watched the priest huddled in the small space he claimed at the prow, checking and rechecking his few belongings. "It'll be a day or two yet, priest," he reminded him. "You'll be seeing nothing but a salty fog for a while."

The following morning was obscured by a damp haze and, as Arne predicted, a thick fog engulfed the fishing banks. Light zephyrs carried the occasional sound of unseen fishermen, but nobody could tell how far the voices had carried, let alone what languages were being spoken.

Standing back at the steerboard and straining to see any vessel in front of them, Arne believed he could make out the silhouette of a lone boat in the distance. He signaled to Stok to have four of the rowers lower the sail. To the rest, he spoke in a hushed tone. "Take us slow ahead. If those are fishermen out there, I don't want us close enough for them to see us. We'll stay obscured in this murky haze."

Arne beckoned to Benji and whispered. "Those fishermen's voices out there tell me this is when we give our passenger a lesson in survival at sea. Grab the gaff, go down in the empty part of the hold and straighten the burred ends off a few rivets. Remove the round iron disk holding them in place so you can knock a plank loose, but save the disks. Let the water seep in and get yourself soaking wet before you come back on deck."

Benji hesitated. "And the priest will be rescued, right?"

"He's the whole reason I'm doing all this, now do what I'm telling you."

Only half the longboat crew was aware of Arne's plan to fake the sinking. He wanted to see real fear from the others. "It'll improve the show," he decided.

Nobody bothered to ask why Benji slipped below, but minutes later, his head re-emerged, dripping wet. He shouted loud enough for all to hear, "Captain. How much water is supposed to be sloshing around in the bilge? I think the whole ocean wants to get in the forward hold." Half of the crewmen pretended to be fearful. Arne enjoyed the real fear emerging on the faces of the rest.

Hiding a smile, he and Stok clomped over to the hatch. "What do you mean, 'it wants to come in'? There's always a little down there, but it's not supposed to *come in*." The aggressive tone of his voice made the situation sound precarious. Benji climbed out of the hatch and stood amidship, dripping wet. Half the crew bellowed for help into the haze and prepared to jump overboard.

"Lower the dory over the side," Arne ordered. "The rest of you men, gather your gear. Prepare to jump."

**

Niels' hand went straight to his neck for the Saint Christopher medallion. He panicked and couldn't control his chattering teeth. He wanted to swear an oath out loud and wasn't ashamed to admit he wished he were back in Aarhus.

Amid the clamor on deck, Stok shouted at Niels. "Priest. Here's a wineskin full of ale. Grab it and climb in the dory."

Niels felt privileged to have been saved, but pangs of guilt swirled through his head.

"You told us you wanted to see Ireland," Stok insisted. "Those are Irish fishermen out there. They're Catholic fishermen who'll scoop you up and deliver you to dry land. You tell them to come back for us after they drop you off."

Using the few Celtic words he learned one evening from a tavern wench, Stok stood at the rail and roared, trying to sound as frightened as he could.

Arne joined him at the railing. "I can't tell what you are trying to say, brother, but you're frightening *me*. If I were the fisherman over there, I'd think this was a pirate attack and scoot off in the other direction as fast as I could." Stok hid a smile and shouted again.

Niels was trembling. He picked up his satchel, carefully inserted his rosary, Tursk's map, and the collection of relics he had secured in the Faroe Islands, and grabbed the wineskin. With a dogged expression on his face, he stepped to where Dag was holding the dory steady, vaulted over the rail on his own and dropped into the center of the unstable little boat.

With his arms stretched out straight to grab the gunwales, he didn't notice his satchel, with all the items he valued, slip off his shoulder. It floated on the water for a moment but immediately darkened as it became wet. Nobody noticed it or the wineskin before they quietly sank.

"Arne," shouted Benji, "Niels can't handle the dory himself. You promised to help him. Pick somebody to get in there with him."

Arne pointed to two of the crewmen, told them to climb in and grab the dory's oars. "When you're rescued, play dumb," he cautioned. "Someday, I don't care how, get yourselves back to Aarhus, let the priest tell the bishop the details of the sinking and how a fisherman saved you. It should be easy to get a boat back to Stavanger from there. There'll be a double crew share waiting for each of you." The two dropped down into the dory and grabbed the oars.

Benji grabbed a bucket to bail the bilge then climbed in farther to pull the plank back into place. To be sure it held, he pounded replacement disks back onto the rivets and burred the points over. The men who knew Arne's plan revealed the deception to the others and assured them their longboat was not going to sink. Trying to mask their amusement, they continued to call out for help.

"Here's more mead," Arne shouted to Niels, and lobbed two full leather flasks.

They landed with a thud in the center of the little boat. "Don't worry, there'll be a fisherman's boat coming for us too. We'll be picked up."

* * *

Within minutes, the three men and their dory drifted alone. The Norsemen continued to row in short bursts when they heard a sound, but in the dense fog, they could see nothing. Niels began reciting the *Ave Maria*, then stopped when he convinced himself he was hearing responses coming back. Several times he shouted, "Over here. Are you Irish? We're here." There was no response.

By nightfall, the two Norsemen stopped bothering to row, and the dory drifted. Niels continued to call out into the empty space surrounding them. "Hello. We're over here," he implored, but there was no response, no indication of any fishing activity. Thinking he should console his companions at one point, he offered a positive assessment, "We three are the fortunate ones, my sons. With no boats anywhere near us, the others must have all drowned by now." Again, he received no response.

"Join me in a loud and determined recitation of the Hail Mary. Surely such a plea will be carried across the waves to a Christian boat crew. Of course, you two remember the words." The two Norsemen said they did not.

"No? Don't worry. I have been taught to project my voice using my diaphragm. If I sing for our lives with enough energy, those Irish fishermen out there will surely hear me through this gloom." The others scoffed at Niels and each picked up a flask, taking long swallows.

After exhausting his repertoire of hymns, Niels dozed fitfully, too frightened and uncomfortable to sleep. His mood improved as the fog lifted and he spotted a rainbow. "See, the light of the Holy Spirit is with us this morning my friends."

The Norse crewmen agreed the sight was worth celebrating and passed a wineskin to Niels, warning him to take tiny sips. Niels tried but couldn't get enough from the bag to wet his lips. "Wasn't there another wineskin?" he asked. It was passed to him, and he discovered it had been sucked dry too. Without reacting, he knelt to begin his day with a prayer, pouring out his favorite incantation, *"Domine Jesu Christe fili dei vivi natus de Maria virgine per quem facta sunt omnia.* There. Something any good Christian fisherman out there will recognize from Sunday Mass."

A disembodied voice floated back. *"Demos gracias a Dios."* It wasn't the response Niels expected, but it was similar to the correct response from altar boys. He froze, stood in the dory with caution and whispered to the other two with him, "See. There *is* a boat out there. And there's a church-going man on board. Yesterday was the darkest of days, but now we're saved. Row to the voice over there."

Within minutes, the dory nudged up against an old and rundown caravel. The crew of four fishermen helped Niels climb aboard while the other two castaways stored the oars and secured the dory to the stern of the tall fishing boat.

Niels tried to explain that their dory was, in fact, a lifeboat set adrift from a foundering Norse trading vessel. "We are the lucky ones," he uttered, half crying with joy. The words caught in his throat, and he crossed himself.

The captain recognized Niels as a priest and explained in a thick Galician accent, "We will take you back to Galicia with us, *Pater*."

Niels recognized the word *Pater*. "At least you recognize me as a priest, and I am most thankful, but one other word troubles me. Did you say Galicia? Isn't Galicia one of the Spanish realms?" Niels was dismayed to think he was now on a boat from the Spanish coast. "Are we going to Spain? I was told you'd be Irish and would deliver us there, if not straight home. My home is in Denmark. Now we will be heading south, the opposite direction and nowhere near Ireland. Have we abandoned my need to return to Denmark?"

The captain didn't understand a word the anxious priest said but scrutinized the well-crafted dory with interest. Niels turned to the two Norsemen, muttering, "I have no idea what this man said or if he'll want to claim our dory as his reward? He'll earn a fine pence if he does, and we'll be stuck wherever he lands us, far from Denmark and with nothing to show."

Niels looked at his companions. Neither appeared the least worried where they might end up or if their dory would be kept by the Galician fisherman. To add to Niels' discomfort, the man's crew continued to set and haul back their herring nets for another two days and nights. The lingering odors of wet wool and herring guts made Niels sick all over again.

Standing back in the stern, trying to inhale as much fresh air as possible, Niels attempted to describe to the captain all the wonderful fishing grounds they had discovered to the west. "Cod is much tastier than your herring, and I know of a place with boatloads over there. It's called *Beothuk*. You should decide to fish there too." Despite the captain's polite smile, Niels was not sure if the captain understood what his passenger was saying.

In harsh tones meant to frighten the priest, the two Norsemen pulled Niels aside and warned him to keep the details of the enormous schools of fish to himself. One poked a finger in his chest. "You're not to tell anybody there is a Vinland, but I'll tell this to you." The man leaned in closer and whispered. "Your young Greenlander friend Benji, he insisted Arne wait until we heard some fishermen before we launched the dory. You'd be swimming right now if Benji hadn't made him agree to save your hide. Just be thankful and say no more to this fisherman."

Niels was chastened by the scolding and kept to himself for several hours.

* * *

On the longboat, Arne checked his compass and pointed off into the fog. "Stok, set a course to the east. We should be near the mouth of the big river leading up to Limerick. It's the best market town in all Ireland and I want to head up there when this fog burns off."

At daybreak, the sky was unusually clear. Arne climbed part way up the serpent's head hoping to spot the peaks of Ireland's ancient coastal range. Following the coast northward, he shouted, "Find a small bay below a monastery on an island. It's on a grassy hilltop. I think I remember the mouth of Limerick's river. There'll be a cove where we'll determine which way the tide is running and wait until the river is high enough to take us upstream. There'll be a fish monger up at the wharf who'll pay us for this load."

Stok gave a belly laugh. "Those Irishmen won't be pleased to see a Norse longboat, but they all enjoy seafood with their ale. And come to think of it," Stok chuckled, "after a month at sea, I'll enjoy their ale too, almost as much as going upstream like a fish ... to spawn."

Chapter 12

Dingle, Ireland

Aideen's return from Limerick had left her gloomy for days. She had collected enough salt for several boat loads of cod but only received modest deliveries from the local boats. If she was to ever succeed, she needed more. Even though her bootleg wine was generating some income; she knew the fishermen didn't like the clumsy process and were reluctant to bring in much more.

"You'll have to persevere," she had told them. "Until I get word from the abbess, I have no idea if she informed the Vicar General in Dublin of my plea for tax relief."

Spending far too much time staring out at the empty horizon, she had yet to have one foreign fisherman offer her any cod. Some whaling vessels did come in for supplies and left with promises to spread word of her plans for a salt fish business, but few offered much optimism, or seemed particularly interested.

Poppie joined her at the end of the wharf and pointed toward the western sky. "Look Aideen. Those seabirds fly low when they head for land. I'm thinking they're running from a squall or some heavy weather coming in from offshore. Maybe it will blow some of your fish our way."

"Sure I wish bad weather on no man, Poppie, but if what you say is approaching, maybe the cod will swim back from wherever they've gone. It would be welcome."

* * *

It wasn't the squall Poppie's seabirds had forecast, and it wasn't welcome. The tail end of a mid-summer cyclone blasted in from the open ocean, whipping the sea outside Dingle Harbor into a turmoil and threatening any vessel coming or going, large or small. The lashing winds continued late into the night, but it was the sound of the winds that unsettled Aideen the most. She curled up trying to ignore the howling gale and driving rain, until a single, huge surge broke right over the wharf and slammed against the walls of her garret. It ripped roofing shingles from the timbers above her, allowing dribbles of rain to trickle down through the rafters.

Unwilling to crawl back under her wet wool blanket, she lit a fire in the brazier and placed leather pails below the worst of the drips. She checked the ceiling and frowned. "I guess this would be the weekly rinse-off I once made fun of."

At dawn she glanced out of her windows at the sea. A humble karve appeared with a handful of men aboard, wallowing recklessly through the swells. "Who would willingly venture into a blow as bad as this?" she exclaimed and watched as the men huddled low in the center of the boat, clutching each other and bailing sporadically.

"Hang on," she yelled and rushed outside. When they drifted close enough, she yelled again, "Here, grab this rope."

The rope missed its mark and the karve slewed sideways, well past the wharf. In the half light of morning Aideen couldn't be sure of where it might end up.

"If you're lucky, you should get tossed up on the beach behind here," she shouted and ran out of the storehouse to the path leading to the cove below. A tidal surge drove the karve ashore, well above the normal surf line and deposited it at the base of a dune. Six exhausted fishermen stepped out and flopped down on the wet, chilly sand, thankful to be alive.

Leaning into the driving rain, Aideen scurried over to the men, hoping for signs of life.

"You're all alright now," she reassured the first bedraggled fisherman. "I've got shelter for you just a short walk up beyond the dunes. We'll get you dried off and warmed up, then see if we can secure your boat."

"*Jeg forstår ikke. Snakke Norsk?*"

"I do speak a little Norse and understand a bit too," she explained. "I learned some of your language from my father's friends as a child, right here on this beach. Let's see if we can get you all up to my garret and dried off. And before your boat is swept back into the sea I'll have Poppie come and secure it for you."

She looked down at their karve and to her surprise, the Norsemen's catch lay secured between the pen boards, still intact. "Those are cod fish you have, and by the look of them, I would say they're in better shape than the lot of you six."

The men eagerly followed her back to the storehouse.

* * *

Aideen ushered the men inside her tiny room, organized cups of warmed mead, then passed around what blankets she could find. When she explained where they had fetched up, none could believe how far they had been blown off track.

"We're from the Faroe Islands," the captain of the fishing karve explained. "The hand lining was good when we set out, and we didn't think the storm was going to be this bad, so we headed out farther and kept at it. Of course, we're glad you found us, but we're too far away to be able to get all our fish back to market. What a waste."

Aideen brightened. "Perhaps there's something we can do with them. I happen to have a barrel or two of salt, so we can put them up right here."

"You don't understand our fishery. We have a contract. Everything we get has to go fresh caught to the buyers' guild in Norway. They pretty much own our islands and all we do there. They'll know if we sell fish somewhere else, but maybe," he wavered a moment. "What if we say our load was swept overboard in the storm? Who's to know you have it now? Mind though, don't be telling anybody."

Aideen was grateful for their gift. It reminded her of what her father had told her once, 'the surf seems to always give back'. Despite being given the catch, she was disheartened to hear how all Faroese fishermen were forced to sell only to contracted buyers. "If anyone asks then I'll say only that they came from the sea around us."

The Faroese crew stayed late the next day to help Aideen salt their fish before they thanked her and sailed for home.

* * *

With sea conditions back to normal, Sean's crew also resumed their trips offshore. Late one morning, he eased his currach back to the wharf. Poppie grabbed the tarred end of a hemp rope and secured the boat to a bollard.

"Good to see you safe home, my son," Poppie said. "But you've not brought in much to show for your day."

"No, no. It was a great day, Poppie." Sean pointed to the wooden cask at his feet. "One mighty hulk of a tall ship passed by offshore this morning. The Spanish captain must have heard of Aideen's offer and agreed to tap a few of his barrels into four smaller casks and trade them to us for the *scadán* we had in the boat. I've left this one cask on top, but the rest are hidden under our nets. The rest of the boats will be in soon with the same. Tell Aideen to come out and see the 'cool drool' she was hoping for."

"This is a nice surprise then. Aideen's busy out back right now sorting the fish she got from the Faroese, but she told me once the first keg has to go to the nunnery in Limerick. So, I'll take this first one and set it right here where the vicar's spies will see it. They'll tell vicar where it's going so he can't charge us tax. After dark, the rest can be fitted up in the rafters to age. It might age with grace like I'm doing."

"Good plan, Poppie" Sean agreed. "At least we won't have to climb ashore with wineskins down our britches. This business is a sight easier to handle when the wine is still in the keg."

"Or in the jar," Poppie laughed.

"Well Poppie, I think these wines are a little too fresh for drinking."

"Most days I'd agree, my boy, but that's usually after I've had too much of it the evening before."

* * *

Aideen was surveying her latest load of fish, now lined up neatly along the wall. *It doesn't look so big in this big room, for sure, but it does have a nice smell to it,* she thought. When she heard the commotion and the delight in Poppie's voice, she stepped out to the wharf.

"Look at what Sean was able to bring in this fine day, Aideen. It'll surely put a fine smile on all the faces down at Bessie's, not to forget your sisters at the nunnery."

Aideen brightened at the sight. "I'll head over there now to find Paddy at his midday meal. He'll be needing to ride this cargo up to Limerick on his next trip. And maybe I'll go along."

Leaving Poppie to look after the rest of the wine, she packed one basket of the Faroese salted fish onto her cart and set off for Bessie's with the first smile she had had on her face in a long time.

"Cousin Bessie, I have a delivery for you," Aideen announced, sitting herself down on a bench in the back. Bessie came over carrying a jar of ale for them both and sat down on a stool next to her.

"I have to thank you cousin. So many new customers from all over County Kerry are stopping in. And without a lie, they're all demanding to try my Balls o' Fish."

"Your what?

"Your salt fish, of course. My man and I have been trimming the pieces we get from you into smaller, bite-sized balls. Serving them now as 'Bessie's Balls o' Fish' we are. Folks gobble them up. And what's better, they're demanding extra jars of my ale and your wine, for washing it all down. What I'm saying is things couldn't be better. My man and me are near run off our feet."

"I'm not sure how many more new fish dishes you should be inventing Bess. My warehouse floor is showing too many vacant spaces, and the gannets are complaining that I don't have any scraps left for them. When I run out there'll be nothing to keep me busy, so since you are 'run off your feet', I might have to come back here for a while and be the help you need so I can eat myself."

"But Aideen, if you have no fish to deliver, I'll have no more customers who walk in asking for it. Without customers, I'll not be needing your help."

Aideen spied Paddy enjoying his lunch and asked Bessie for a mug of ale to carry over to his table. He was pleased to see her but asked, "Well, thank you, but what am I going to do to deserve this?"

Aideen explained her plan as he slurped from his ample bowl of pottage. "I have promises to keep. First, there are the kegs of wine I pledged to the nunnery in Limerick. If you agree to deliver them for me, I might go along for the trip. During that big storm a while ago, I met some fishermen from the

Faroe Islands. They say they have lots of fish, but they aren't allowed to sell any to me. I promised to keep Bessie in salt cod, but my supply is running so low I won't be able to fill the demand for her Balls o' Fish much longer. There has to be a way to work around their rules and find a peddler in Limerick who can sell to me."

Paddy listened to Aideen's plans, less interested in the details than having a paying passenger for a trip upriver, and a single woman to boot.

"Well then, you're in luck, dearie. Tomorrow afternoon is the next good tide. I'll come by and collect the kegs early tomorrow, and you be ready to climb aboard right after I have my lunch here. Who knows? Maybe there's a peddler who knows somebody who is allowed to sell Faroese fish." He slid his hand over and gently touched Aideen's arm. "And it'll be nice to have you on the boat again, Aideen."

Chapter 13

Limerick, Ireland

Approaching Limerick, Arne was pleased to see the harbor crowded with boats and busy with people.

"See those huge castle walls. There must be hordes of hungry folks inside. For sure, some of them have to be hungry for a load of stockfish."

"And they all get thirsty too," quipped Stok with a smile. "They must have a good selection of alehouses in this town. Let's find out where they drink."

"We also need to find a trustworthy carpenter here in Limerick to patch our hull properly. I want it seaworthy because we're in the fish business now and business is good."

The longboat threaded its way through the shallows and into the boat basin where a forest of tall masts swayed back and forth. Arne spotted a group of men unloading barrows of turnips and shouted, "Where do honest fishermen sell a load of fine stockfish?" A few heads rotated but ignored the harsh sound of the Norse language. Arne tried again. "Is there anyone who'll tell me where there's a buyer who wants to make himself a profit on this load of fish?"

"Maybe," one voice spoke up. "You mean Viking fish, but if it's a decent quality, I'll give you a fair price."

Arne ordered the boat secured and he jumped off to dicker with the vendor until a price was struck. When the unloading got underway, Stok chimed in. "Peddler, tell us where to find a full pitcher of your Irish brew."

"All the public houses in this part of the city are run by the Normans and you'll not be too welcome, but there's a busy alehouse across the River Abbey where us folks get along with you Norsemen. Take the Baal's Bridge," the man explained and pointed.

Arne told Olaf and six others to finish the transfer of the crates of fish to the vendor.

"Why me?" Olaf grumbled. "Let the smart-arse Greenlander and his friend do some real work for once."

Arne hated defiance, but he needed Olaf's cooperation. "The peddler needs a few strong men like you to help. You'll be staying with Stok and me until we've settled up. After we're done, the ale is on us." With the dispute settled, the rest of the men headed across the river for the bridge. "I hope they are open for very thirsty drinkers," Benji laughed.

* * *

In the bright afternoon light, Stok watched the final crate of Greenland's fish being placed onto the peddler's carts, all the while entertaining the driver with outrageous exaggerations of his past exploits. The peddler gave Stok a full purse in payment, along with a word of advice.

"Not so many Norse families buy hard-dried fish like this anymore. I'm not complaining. These fish of yours are fat and will sell better than the fish from the Faroe Islands, but I'd be giving you more for it if it were salted. These days, people want salted fish because it's easy to keep. When you're ready to eat, all you need to do is soak it overnight."

The difference was lost on Stok. "Fish is fish, I say." He bounced the full pouch in his hand and slapped the man on the shoulder, "until it turns to silver."

With the deal complete, Stok's mind turned to drinking some of the profits and a chance to chat with a barmaid or two. He went to find Arne, who was inspecting the right side of the longboat's hull where the carpenter was completing the patch.

Stok wanted to celebrate the trip's profits and walked over to his brother. "We've pocketed a bursting pouch for this load." He pointed back to the peddler. "Dealing with him is better than searching high and low for walrus ivory to sell to the church. Since we're back safe and dry now, let's go spend these coins. There'll be lots of time before the next tide out of here."

The peddler butted in. "I've got some more free advice for you men. I've been working these alleyways all my life. My pa peddled fish long before I came along. All told, we've carried a lot of things to this market, including ivory and I'm here to tell you there's no more money to be made in walrus tusks. There's a lumbering gray beast living somewhere far to the east. It's called an elephant, and its tusks are pure white. The Pope says they polish up better than walrus tusks. His cardinals are already signing deals with traders who bring great numbers of them in cheaply on traders' caravans. And those cardinals are telling all their bishops not to buy walrus tusks anymore. You two should keep filling this longboat with codfish, salted fish, that is."

Stok turned to Arne for confirmation. "Well, after faking our disappearance, we won't be able to deal with the church anyway. So no more tusks. We'll keep the boat for ourselves and fill it with fish, right?"

"Maybe, but we need a plan before we do this all over again. If we are going back to the far edge of the world, I don't want to get pounded by another storm on the way."

* * *

Surfing the upstream current and gliding right up into Limerick's busy waterfront, the ride in Paddy's curragh seemed far less distressing than the first time Aideen made the trip. At the wharf, Paddy peered through the maze of vessels tied alongside and recognized the shape of a Viking longboat.

"It could be a good sign, Aideen. The crew on that boat might have come from the Faroe Islands. Maybe they'll know where to find Norse fish peddlers."

"You may be right there, but it doesn't look like anybody's around. Would you have an idea where the crew has gone off to?"

"If I was to guess, I'd say they'll be across the Abbey in Irishtown and likely already at the alehouse. The barmaid is named Tilly, and she understands a bit of the Norsemen's language. I'll be off to the nunnery with these casks then. Mind though, I'll be looking to leave when the river is as high as the top of the bridge pillars there. If you plan on coming home with me, don't get turned around in Irishtown or waylaid at Tilly's."

Aideen gave a slight flick of her hand. "Don't be worrying for me, Paddy. I've been here before."

* * *

It didn't take Benji and Dag long to locate the alehouse. "Dag, look up there." Benji pointed up the alley at a two-story, wattle and daub building with an oversized hand-carved mug hanging low over the entrance. "That must be the place. I should have a big wooden mug for my brewery back home, too." As they walked closer, they both got a whiff of malt and yeast coming from the building. Benji announced with delight, "This is where I'll have my first drink of ale in a foreign land."

The two stepped inside, but neither could spot an unoccupied table, so Benji approached a group of four young men and introduced himself. Tapping his chest, he said, "I'm Benji Leifsson. My friend here is Dag. We're fishermen off the longboat tied up near the castle. Can we join you?"

The locals didn't understand what Benji was saying, but a lad named Fionn recognized Benji was speaking Norse and tried to answer. "My grandfather was from Norway. He taught me a bit of Norse so maybe I can help." He scooched his three friends down the bench to make room for the two strangers. "This might be some fun for us."

"So you two are from Norway I'm betting," Fionn said in his broken Norse." Maybe you're off the longboat I saw at the wharf. Are you innocent fishermen visiting our town or maybe you're the last of the Viking hordes?"

Dag sat up. "We are both of those. I've done some raiding and not ashamed of it, but we're fishermen now." He pulled a thick section of stockfish out of his sack, dropping it on the table with a thud. "Want a bite?"

Dag drew his double-edged dagger and set it on the table next to the dried fish. Fionn and the others arched back.

"This blade will have to do because I left my ax on the boat." Chuckling at the reaction around the table, he carved six strips off the dried fish and dealt them around, leaving deep scratches in the table. Fionn translated what Dag was saying, hoping to settle the nerves of his Irish tablemates. The others declined the fish until the barmaid arrived with a tray of ale. She asked for a bite and insisted the fish went down well with her ale. Following her lead, the men each tried a bite.

Hoping to divert attention from the long knife on the table, Benji spoke up. "Not all of us are raiders, or fishermen for that matter. My family and their families before them owned farms in Iceland but left for Greenland after volcanic ash smothered their crops."

Fionn spoke up. "Greenland, you say? I didn't think anybody *lived* in Greenland. What in the name of God's bones is there to do there except maybe eat walrus?"

Benji became more serious. "Small villages of farmers have been there for generations. I live in the Eastern Settlement. It's up a long fjord — Eiriksfjord, the one discovered by Eirik the Red himself."

Fionn translated for the others, still trying to lighten the conversation and asked, "Why didn't the old Viking call the country Eiriksland, after himself, instead of calling it

Greenland. Was the name a joke on your ancestors to trick them into going to the land of snow and ice?"

Benji had never questioned the name. "No, no. Greenland is the proper name. It is green, well, half the year it is. The first pioneers arrived to hunt the walrus and collect their ivory, but you're right — can you imagine eating walrus and fish, meal after meal? They needed farmers to raise small animals for meat and to plant crops, so my grandparents agreed to go. The king gave them free land to plant barley and oats during the long summers. And after all these years, they still love it. It's nice in the summer. Harvests there are better than any in Denmark. We always peel off our clothes and jump into the pond to cool off after haymaking."

Fionn's tablemates butted in. "What did he say, Fionn? What did he say?" Fionn played along. "Mr. Benji Laugh-some, they want to know what you do in the winter, when it is so long and dark?"

A chuckle rose from the table as Fionn mispronounced Benji's last name on purpose. Benji was enjoying the local brew and didn't understand he was being mocked.

"You're right. I don't mind watching the crops grow all summer, but sitting in the dark waiting to do it all again the next spring isn't much fun." He changed the subject. "That's why I joined up with Dag's longboat — to visit interesting places, like here." He carved another strip off the dried fish. "Fishing days are long, but the catches in Vinland are great. When we were there, we couldn't put a hook in the water but it had a fish on it, sometimes two. If they are not big enough, you scoop the cheeks out to boil later, toss the rest over and hook a fatter one."

"Now you're telling us Viking fish have cheeks."

The barmaid arrived with another tray of ale as Benji proudly explained, "They do have cheeks." He took a fresh swig and puffed out one cheek. "It's the sweetest parts of the fish." Ale dribbled down his tunic and the group doubled over laughing at him. Tilly too.

* * *

After traipsing through cobbled alleys and getting unhelpful suggestions when she asked for information on fish peddlers, Aideen remembered she was soon supposed to meet Paddy for her ride back down the river. Not knowing where to turn, she remembered Sister Enda and the offer she made during her previous visit to Limerick. 'You're welcome to come back here, any time of the day or night'. Aideen looked into the night sky to be sure she could locate the nunnery next to the cathedral's tall spire and find her way back to Englishtown, then renewed her search for a fish vendor in Irishtown.

On a bustling lane packed tightly with timber-framed cottages, she spotted a public house. The back of the building was built up on stilts and extended over the banks of the River Abbey. At the front, an oversize wooden mug swung from the overhanging second floor rafters. The gravelly voice of a barmaid resounded from within.

Three men stumbled out into the alleyway as a corn broom was swung at their heels. "You two souses carry your drunken friend home to his wife and put him to bed. And don't come back until the works of you are sobered up," she demanded.

Aideen stepped back, careful to avoid the wobbling trio and the barmaid's broom. Crossing the threshold, she noted the alehouse's one main room was much larger than Bessie's, but the gregarious atmosphere and the familiar aroma of stale ale made it seem as comfortable as her cousin's brewhouse. *Inns and public houses don't change much,* she thought.

Once inside, Aideen waved a hand to the barmaid and called out, "I'll have a mug of that fine ale I'm smelling, if you please." A small table stood empty across the room, and she hurried over to stand there before another patron claimed it. Two steps before she got to it, she tripped on a tilted floorboard, landing hard on her right elbow. She hopped back up, hoping nobody noticed, but flushed a brighter shade of red than her hair. None of the other patrons turned a head.

"This floor of yours is all wiffled," she protested, when her ale arrived.

The tall blonde serving wench laughed at Aideen's description of the floor and ignored her obvious embarrassment. "Folks call me Tilly, and I'll be pleased to tell you the story. This building was extended over the riverbank to give us more room for customers, but we have to deal with a minor flood at the full moon. On those days, the river bubbles up through the boards. It's worse when the high tide time is the same as the busy evening drinking time. The good thing is it rinses the floor on the way back out to the street, but the boards warp over time. You get used to it."

Aideen appreciated the honest explanation, gripped a stool and carefully sat. She could hear the water lapping the pilings underneath her. "It reminds me of my own wharf," she told Tilly, "but I might come up with a better solution at my property than you have fashioned here."

Tilly placed a wooden mug, full to the brim, on the table. It made a dull clunk, and the top layer of foam slopped over. "Sure, this'll make you right, *a Chara*," she said, adding the local expression for 'my fine friend'. She turned to deliver ale to the next table but called back to Aideen over her shoulder, "No charge to you for the first one."

Aideen hardly heard Tilly over the loud chatter of the young fishermen at the table next to her, enjoying their jokes and taking no notice of others around. She was unable to understand what was being said, so she massaged her elbow and watched the attractive lad with the curly blond hair. After a long swallow she set her half-empty mug down on the table. Feeling better, she gave a nonchalant scan of the room and recognized Paddy standing against the far wall. Their eyes met, she waved, and he ambled over.

* * *

"Paddy! I expected you to be gone downriver by now. I'm happy to have your company though. Please join me," Aideen said.

Paddy snatched an unoccupied stool from a neighboring table and sat. "After I left you to deliver the wine, I remembered you needed to find some fish. In this city, as many deals happen at an alehouse as at a vendor's stall, so it made me think I should drop in here and see what I might see. As it turns out, I do have a tale to tell you." Pointing to her drink, he added, "But first, perhaps you'll be kind enough to order me a wee one of those."

"Surely. I'll ask Tilly for another mug ... on me."

Before the ale arrived, the loud banter at the table next to them took their attention. Paddy pointed to the handsome young man gnawing on the dried fish — the same lad Aideen had been watching.

"The tall lad there is Norse," Paddy explained. "He claims to be a fisherman. The others at the table don't understand much of what he is saying so he thinks if he talks louder they'll understand better. Anyway, the talk is largely pointless, but he is from Greenland and he's off the Norse longboat you saw at the wharf. If you want fish, you should go over there and buy a round for him and his new friends."

Aideen took a minute to recall some of the words of Norse she picked up as a girl and hoped the lads would understand her. Stepping carefully over the uneven boards, she headed straight for Benji. He lowered his glance when he saw her approach, but not before Aideen saw his piercing blue eyes. She pointed to the sliced pieces of stockfish on the table and spoke directly to him in her halting Norse.

"The fish there looks dry as a board, but I bet Tilly's brew helps it slide down right fine. May I offer each of you another mug full of the local drink in return for information on the source of your codfish?"

Benji gave her a look like he didn't understand what she was trying to say, but Fionn gave her a half-sober grin. He was used to hearing Norse words tinged with an

Irish lilt and spoke up. "We'll gladly agree to another round for the table, but these two lads will tell you there's no fresh cod near here. Certainly not fat ones like the one there. They've moved north, or west, maybe as far as the land of Mr. Laugh-some here."

Aideen didn't want to speak with Fionn. "So could you ask this attractive Norseman for me if he knows of a steady source, perhaps from one of the Norse colonies."

"No problem. According to this handsome Greenlander you've fixed your gaze upon, he has what you need. He already told us it comes from a broad bay somewhere beyond Greenland and it'll be dried hard as a board by the time you see it."

Fionn picked up Dag's sharp blade, chipped a wedge off the flake and clamped it between his teeth. "This is what they deal in. You can hardly rip it apart with a Viking's ax."

"Stockfish," Aideen responded, her disappointment obvious. "My fishermen can't come in with the likes of that and claim it's fresh caught. Enjoy the drink boys, but tell him to keep his dried fish." She sauntered back to Paddy.

"The fish will have to wait until I speak to somebody who knows the difference between a treat from the sea and what's been dried hard as the floor we're walking on. You and I may have to be content with importing wine."

Aideen got ready to leave, and Paddy downed the rest of his ale.

"Well, the bad news is we've missed the outgoing tide, Aideen, but the better news is we'll be fine all snugged up close under my sail cloth for the night. We'll keep the chill off each other."

"An appealing offer to be sure, but I'll be getting myself back to the nunnery. They offered me a place to lay my head."

"More's the pity. It's not the first night for me to be sleeping alone alongside the Limerick pier, though. I'll be expecting you to come by tomorrow morning, then."

Aideen bid Paddy farewell and turned to wave goodbye to the table of lads, hoping to meet Benji's eye. She did, but when their eyes locked, he quickly dropped his gaze.

Chapter 14

Limerick, Ireland

The Gunnarsson brothers burst into Tilly's alehouse, followed by the rest of their crew. Stok surveyed the room and jostled his way over to an empty table. Tilly appeared carrying two pitchers and a tray of mugs perched on her fingertips. She put the tray down, filled several mugs to overflowing and said, "You men look as if you've been adrift for a while. Am I right?"

Stok had barely gulped down a huge mouthful, so he gave her his best toothiest smile, causing ale to dribble from both corners of his mouth. Tilly kept talking, this time in broken Norse. "Are you off the same longboat as Benji and Dag, the longboat unloading Viking fish?"

Stok couldn't believe his luck. The wench was chatting *him* up and in his own language. He swallowed again and his face brightened, "You're right. And we're going straight back, but before we leave we'll down this ale here and you can bring us another round."

Tilly opened her mouth to speak, but Arne brushed her off. "We have things to talk over here."

"Suit yourselves. I'll keep watch and be back when your mugs are drained."

Arne turned to Stok and asked, "Did you notice the streets in this town? They are full of ordinary men and women, not Catholics who eat fish because their priest says so. These folks are hungry. There'll be a strong demand for our fish right here."

Stok was trying to follow Tilly's movements around the room, but said, "Arne, I don't remember you being so enthusiastic about anything as honest as this before."

"And here's the best part," Arne added. "Remember what those fishermen in the Faroe Islands told us? They get all the fish they want, and they leave lots more swimming because there aren't nearly enough people on their islands to eat it all. And Baldr owes us a favor for all the free supplies we left there, right? We'll go there next and have him fish a boatload for us."

Arne took a gulp and went right on with his plan. "You saw all these poor peasants around here. Once we put a fish in their hands, they'll be begging us for more. We'll keep running back and forth and never have to go all the way to Greenland again."

Stok wasn't thinking of fish. "Give me a minute. I need another drink." He waved his hand and went over to where Tilly was standing.

Returning with Stok and another tray of ale, Tilly jumped right into their conversation. "I don't miss too much whether it's in Irish, Norse, or the Normans' English. And you'll be interested in another bit of fish talk I happened to hear."

Arne gave her a blank stare.

"Right ... don't listen to me, but there was a fair maid from Dingle sitting over in the corner a while ago looking for fish. You should have a listen to her. A lovely dear she was. Spoke a little Norse too, she did. She left not long ago, and I reckon she'll be over at the nunnery by now, curled up in her sparse cell for the night and accepting a full belly on the morrow. She was looking for fish though. If that's what you're trading, I'd be over the bridge, knocking on the gates of the nunnery first thing. Ask for Aideen."

"There's some sense in what you say," Stok decided. "Fetch us another pitcher, and before we leave tomorrow, we'll see who's who in the nunnery."

Tilly gave Arne a self-assured smile and winked at Stok, who was beginning to understand Arne's enthusiasm for trading fish between the Faroe Islands and Ireland.

"Think of it, Arne, those islands are only a short sail from here. Two days there and the same back. In six days or less, I could be sitting on this bench again with Tilly beside me, the prettiest of all tavern wenches. I'll have a better chance to tell her my stories then."

* * *

A loud clatter outside the nunnery's big gate carried into the dining hall. The abbess signaled for a postulant to check who was there. She returned and whispered to Sister Enda. "Two frightful Norsemen are lurking outside, Sister. They've demanded to speak to your guest. What should I say?" Enda leaned over to tell Aideen. A ripple of disquiet echoed the length of the dining table. Two rows of veiled heads swiveled in Aideen's direction.

Aideen hoped the visitors at the gate might be the Norse lads she spoke with at Tilly's, but those two could hardly be described as 'frightful'. "Who could be asking after me and here, of all places? Prithee, Sisters, forgive me while I deal with this intrusion."

She stepped through the vestry and into the cloister. Gruff sniggers came from the other side of the thick oak doors, and a voice joked, "One single blow from my ax on these rusty bog iron hinges and we'd be in." Aideen gritted her teeth, determined to show no fear, and slid open the tiny peephole.

"Hello. I am Aideen. What is it you think I can do for you?"

Stok bent low, trying to peer back through the opening. "Last night we heard you wanted to buy fresh fish. You want to know what we do now? We sell fish."

Listening to their accents, Aideen realized the men must be off the Norse longboat tied up near Paddy's currach. "That's not what I was expecting to hear at the gates of a convent," she admitted.

"We didn't expect to be talking to a fishwife through a nuns' hole either, but since we are, open up and we'll see who's talking to who."

"I'll be out by and by. Give me a moment to settle up with the good sisters, and we'll talk on our way back to the docks." She bolted the little hatch shut and hurried back to beg forgiveness for the disruption.

"Begging you kindly, Reverend Mother. I hadn't expected my interest in trading fish to cause such an upheaval at this time your family reserves for peace."

The abbess nodded. " You have more than earned our forgiveness child, what with the wine keg delivered late yesterday. My great regret is I have yet to hear from the Vicar General in Dublin regarding your plea for the import tax to be lifted."

"It would be a great help if you could press the man, I cannot stress enough the difficulties we endured smuggling this very wine in from offshore. The Vicar General must act, or we will have to abandon the effort." Aideen knew her comments might be received as strident, but there was no option now. "And please ask the startled postulant who responded at the gate earlier to forgive me."

"***Go dté tú slán***. May you go in health."

* * *

Arne quick-marched through the crowded alleyways of Englishtown. Aideen and Stok struggled to keep up. Hoping to slow their pace, Aideen asked, "How did you know what I was looking for and where to find me? You certainly took the nunnery by surprise."

Stok stopped walking, looking forward to a chance to chat up Aideen. "The pretty tavern wench named Tilly was enjoying my stories and —."

Arne interrupted. "Keep moving. We don't want the longboat to fetch up on the river bottom before we see Scattery."

Aideen ignored Arne and turned to Stok. "Ah, Tilly the barmaid told you."

Arne interrupted again. "You want fresh cod, and we happen to have a contact in the Faroe Islands. By the way, where will we be delivering this fish? And how will you be paying? Silver is best."

"This is where the deal gets a little complicated. I live in Dingle, and I do want fresh fish, but I don't want you to come right into shore with it. And there's another thing. You won't be paid straight away either."

"So, no place to unload and no reward when we do. Is there anything in this scheme to make us happy?"

"The months of summer and autumn each have a full moon. On the first night of the month's full moon, you make a deal for a load of fresh fish with your Norse friends in the Faroe Islands. When you're loaded, you'll head south, splitting and dressing the load along the way. By the following night, my fishermen will cross paths with you a few leagues offshore. They'll take your cleaned fish at sea, over-the-side."

"Over-the-side?" Arne was ready to walk away. "You want us to load our fish onto other boats and watch them carry it away?"

"To sidestep the local vicar and his tax, the transfer has to happen out at sea. A man called Sean, plus a few other boats with him, will take your fish and deliver it to me. For the first load, Sean's men will be packing your fish with my salt and landing it as they do their own. You come in and see me afterwards, and I'll have your payment. Next trip, you'll be doing the salting yourself on the way in. I'll give you the salt you'll be needing. Mind, though, all this must happen outside the entrance to Dingle Bay."

Stok butted into the conversation.

"Arne, there is a nunnery in Dingle. An old Viking told me once he had an unforgettable raid there. If we're going to be there, we might ...."

Arne frowned at his brother. "I've told you before, if you think you're fit enough to go back raiding, you're wrong. Get your arse down to the boat and see the crew is ready to shove off as soon as I get there."

Aideen was surprised by Arne's aggressive response. His reaction and Stok's comments reminded her of how this pair of battle-scarred Vikings had unnerved her back at the convent's gate. She decided to be wary and not expect an honest answer from either one. On the other hand, she wondered if doing business with these savvy operators might teach her a few tricks for future dealings with Vicar Maurice.

At the dock, she left Arne and walked over to Paddy's currach. He was waiting there, and she could see he was impatient to push out into the stream. "River's a runnin', Mistress Aideen," he shouted. "We got to ride it."

Aideen was glad her boatman gave her an excuse to hurry the deal with Arne. "We're done here, Paddy," she said and turned to Arne, "Do we have an agreement?"

"There is one thing. You said you would give us salt. Last night, the fish peddler told us the same thing. Why salt?"

"Ah," Aideen explained, "For one thing, this is a damp country. A layer of salt cures the fish faster. And for a better reason, who doesn't like the taste of salt? It's what we always do with our cabbage and pork. Why not fish?"

Arne gave an unconcerned flip of his hand. "You do what you want with the fish. We don't care after it's delivered."

Aideen climbed aboard Paddy's currach and shouted back to Arne, "See you in Dingle at the next full moon." She didn't say she hoped to see the attractive young member of Arne's crew.

"Benji," she said aloud to be certain to remember. "His name is Benji."

* * *

Arne jumped onto the longboat. "There's a whole load of fish swimming around the Faroe Islands for us to collect for a past debt. The fishermen haven't been told we're coming yet, but after we drink enough of their ale and the moon turns full, the old troll Baldr will give us what he agreed. We'll fill this boat with his fish and continue right back here to Ireland."

"We're not heading back home, Arne?" Benji asked.

"We *are* home, my young Greenlander, home on the sea."

"But we're going back to *my* home after. Right?"

Arne's response was brusque. "Probably not. Get your oars wet. We're bound for the Faroe Islands and four weeks of waiting around until we get what Baldr promised for us. And you should be happy. You don't have to row across more stormy seas."

Chapter 15

A Coruña, northern Spain

Two long days at the mercy of the ocean swells left Niels wondering how much more dismal his life could become. When the captain pointed to the tall Roman lighthouse at the entrance to A Coruña's harbor and tried to explain they had arrived, the bedraggled priest uttered a simple "It's not my home, but God be praised. What more will happen to me now?"

Coming alongside the busy wharf, youngsters grabbed for boat's lines hoping for a few pence in return for securing them to the bollards. Matronly women in long black frocks crowded close, hoping to be tossed a few fish for their family's noon meal. The captain graciously agreed, beaming with pride as he showed off his full load of herring to waiting fishmongers.

Tied up at last, Niels and his two companions couldn't get off the grubby boat fast enough. Niels offered a cool *'Pax vobiscum'* to the captain, thinking *these robes will carry the stink of his fish for weeks*. Seconds later, a pair of gulls glided low over his head. One released a glob of yellowish green slime that glanced off his left shoulder and splat on the ground in front of him. He didn't bother to wipe it off.

For the moment, he crossed himself, tried to pat wrinkles out of his soiled robe, stepped over the splat of slime, and knelt to kiss the rocky soil. "I have been far too long at sea. I vow never to leave the solid ground of our Mother Earth again."

His two Norse companions stepped around him, "You're right, priest. You won't be going back on a boat because this is where we begin our walk — over dry land." The two turned left to walk along the coastal road, heading for the Kingdom of the Franks.

Rising quickly to his feet, Niels pleaded, "Wait. Don't you want me to accompany you?" Neither responded.

"No? Oh well, God be with you."

* * *

Niels watched the two men until they were out of sight. He had no idea what else to do. *Bless me. I know not where I am, nor anybody here.* He struggled for a moment and then looked to the crowd around him. "No, I'm not alone," he said aloud. "All these people are God's children." He searched the skyline for a tall spire, but not seeing any religious structure at all, he asked in Latin, "Where is the church in your village?"

A young lad responded in an altar boy's mixture of Latin and Galician, flicking his thumb back behind him. "Are you looking for the Christian hermitage perched high in those hills, *Pater*?"

Again, Niels understood nothing but the word *Pater*, but it was enough to tell him the boy recognized him as a priest. "Thanks be to you, my son. I'm thinking you have pointed to a nearby place of worship. It's odd though, having your town's church located in a direction away from the village center." Nonetheless, he gave the boy the sign of the cross and headed off in that direction.

"I have no idea how far I must trek, but I'm on dry land, in a Christian country, and I'm beginning to feel the peace I have craved for a long time."

The oddly well-worn path was easy to follow, and in an hour, he stood in front of a humble monastery encircled by a hedge of holly and rows of grape vines. At the gate, he was met by a chubby little friar in a coarse woolen robe and sandals. Niels was embarrassed by his own appearance but hoped he would still be received as an ordained priest. He spoke to the man in Latin, thinking it might give him the needed legitimacy.

"Locating your sanctuary gives me such peace," he confided to the hooded figure. "There's been nobody to hear my confession for many weeks."

The friar was enjoying this chance to practice his Latin and responded as best he could.

"Father, as but a brother, I am not qualified to hear your confession, but if our humble commune gives you peace, you will be delighted to learn we are in the archdioceses of the Grand Cathedral of Santiago de Compostela. And here at our hermitage you are but a further two or three days' walk along the route known as *El Camiño*." Niels felt his heart leap with joy. He couldn't believe this turn of fortune.

"Am I in the vicinity of The Way of Saint James? After Jerusalem and Rome, The Way is the most celebrated pilgrimage in the world. Would I be shirking my duty to return home if I continued on to Santiago de Compostela? What if Bishop Olsen was challenged with the same choices?" As he struggled with his dilemma, Niels was unaware of the friar silently watching him.

"There is no question. This opportunity was Heaven-sent. At sea, I was with men who shared my culture but spurned my faith. Now, I have a chance to embark as a pilgrim to a shrine where my faith is shared as a revered gift. I must seize this moment."

Niels was offered a monk's cloak and invited to stay the night in the stables. He politely refused the hooded robe of harsh woolen fiber, preferring to stay dressed as a priest, despite the stink of long dead fish and the odor of many weeks without a bath.

Walking over to the small outbuilding, he met a band of travelers following their own pilgrimage along The Way. They balked at his appearance and the odor trailing him, but pointed to a stall he could use for the night and shared their fresh bedding straw with him.

Giving sincere thanks to his host when he woke in the morning, Niels rinsed off as best he could and knelt at the friar's unpretentious grotto to say a final prayer. Kneeling at the hand-carved altar, he shuddered when he noticed the adornments were carved from walrus tusks.

"Give me the strength to forget how these adornments come into our hands," he pleaded, and left the hermitage through the rows of grape vines to join the ready band of pilgrims.

* * *

On the road to the cathedral, Niels and the others became close companions. Their mutual understanding of the Christian liturgy allowed him to offer the sacraments and to lead the nightly vespers. When they tried to obscure their sins by confessing in their own vernacular, he would tease them, saying, "I should warn you, I don't understand your regional dialects of Spanish, but the Lord does." He continued to offer them comfort and absolution with a smile.

After three days following a mountain track pounded smooth by the leather boots of many pilgrims and two nights sleeping rough at the side of the trail, the group strolled into Santiago and the crowded passageways meandering between rows of two- and three-story wooden structures. Turning a corner and climbing a slight incline, they found themselves in a grand plaza face-to-face with the grand cathedral.

The two soaring pinnacles stopped Niels in his tracks. "This place is beyond imagining. See how the white spires stretch up through the deep blue of the sky to touch Heaven itself." He looped his dingy embroidered stole over his head, began to weep at the sight of the cathedral's ornate carved façade and wandered away from his traveling companions. Completely forgetting his intention to make a confession, he instead rehearsed the proper Latin phrases he would need to start a conversation and searched the crowd for the first person he saw in clerical garb.

"Will you introduce me to your archbishop?" he asked a hesitant young nun in an unfamiliar white habit. "Through him, I must send a message to Aarhus, in Denmark, where I am from. I have been on a critical errand for the longest time and have not been able to report back on my whereabouts and soon to return." The nun was frightened by Niels' appearance and embarrassed by her lack of Latin. Clutching the crucifix hanging around her neck, she pointed to an edifice she knew to be a glebe house across the plaza. From there, he was directed to the archbishop's office in the chancery and invited to sit and wait for an audience. Niels fidgeted and couldn't stop marveling at the grandeur of his surroundings.

Archbishop Pedro Mateo welcomed Niels and escorted him to the cathedral's library, where he was introduced to a scholar who understood a little Norse. Niels was appreciative of the opportunity to explain his situation and astounded by all the tomes and hand-copied antiquities assembled throughout the room. He pictured how perfect the location would be for him to stay while he committed a full accounting of his voyage to parchment.

"My brothers, before I ask for help in organizing my return, could I solicit your assistance in an important matter? Might I have sheets of parchment or vellum, ink and a quill? And, if I may, the assistance of a clerical illuminator for the time he will need to illustrate my text? The church in Aarhus would indeed be grateful."

The head librarian debated how he might be able to help this newcomer with such a specific and expensive request. "There is a limit to what we are able to offer you, my son. Travelers arrive from many lands at our open gates. We have an endless job, dealing with each sincere request for assistance. It keeps us and all the resources of the church fully engaged."

"But the specifics of my voyage are of great consequence to my parish in Denmark. The wealth of fish I have seen in Vinland will feed all the new Christians at home. My eventual return and the value of my information must not be underestimated."

After a short conference, the library's scholars relented. "Father Niels, the provision of fish during Fast Days is a wonderful thing. There is perhaps space in our scriptorium for you to begin writing and illuminating your report. Once you have completed your work, we will place the manuscript in dry storage until a fellow Dane may be able to carry it back for you. For your other request, we ask you to reconsider your plans to leave us afterwards. As an ordained priest, you are more valuable to the church right here. In return, we can offer your own cell and meals in our glebe house. If nothing else, there'll be a fresh priest's cassock and a clean shave for your tonsure. You're a fright."

Niels stood still, unable to decide between his obligation to return to Aarhus and the opportunity to play a role at this important shrine. The librarian who spoke some Norse restated their plea. "Father Niels, won't you consider staying with us and helping with the flock of daily arrivals?"

The events of the past month ran through his mind, and he responded without further hesitation, "I must remember to be mindful of others' needs. In return for the welcome you have shown me, I am honored to work with you in this eminent basilica." To himself he added, *At least until I complete my manuscript.*

* * *

Each evening, after his work with the crowds of visitors, Niels made notes from memory of his time in Vinland, always emphasizing the amazing fishery in Notre Dame Bay. "Vinland is a treasure with unlimited fish, and I must add an untold number of unbaptized souls who live their lives there without the benefit of the Good Word." He wrote about his misgivings as well. "Some of this is information told to me, and I admit there may be unintended errors or perhaps misleading facts."

In order to replace the map Tursk drew for him in Greenland, he requested sketches of the islands in the bay be prepared under his direction and placed in the margins, alongside renderings of the clothing worn by the young boys he encountered. "With these images, I hope many more adventurers will be encouraged to follow, in spite of the warning from those frightful Norsemen who survived with me." When Niels had finished the manuscript, he presented it to the head librarian.

The scholar reviewed the collected sheets, commented on the excellent quality of the illuminations and placed them in a secure storage cabinet. "Father Niels, my decision to encourage you to preserve information on the location of such an expansive fishery in an uncharted land might have been valid to a fault, but I am reluctant to accept the detail therein. My cousin is a fisherman, and I must tell you, he has difficulty catching one fish on a line. Certainly none as large as you describe. I suspect the Viking heathens misled you when they claimed they filled their boat in a single day."

Niels was immediately defensive. "Regarding the size, I did not exaggerate. As to their numbers, I swear they could not be counted. There was literally no end to them. Catholics all over the world could be fed with all the *Beothuk* — the name the Forest People called it. And in the future, those who come upon Notre Dame Bay will be welcomed by those who live there. They will be granted the opportunity to gather enough fish to feed the masses. At the same time, they could baptize the residents so they may discover God's love."

"In truth, even if what you say is true, nobody will believe such stories. They will discount the whole of your manuscript and besmirch this library where it was written. In my professional opinion, you should rewrite the text with a more realistic accounting of your observations. I mean, if you have the time after you discharge your many duties here."

Niels was crestfallen. He spied his handiwork lying on the floor and asked, "Am I being asked to stay on here and change this document?"

"Neither. You needn't change the wording if, in your mind, it represents your experience. Nor are you obliged to stay. On the other hand, we have grown fond of you. Perhaps you will consider remaining here with us, at least for the short term. You understand that for us priests, duty lies the world over. This institution is a blessed calling. We must hear and respond to the needs of each devout soul who comes to us at any time of the day and night, however impossible it may become for us. There could be added benefit for your commitment. We will offer you tutorials in the Normans' English and Gaelic, plus the local Spanish dialects. Many Franks come to our shrine too. You will receive instruction in a number of their dialects, plus Basque, although you may have difficulty with that last one."

As an ordained priest, Niels understood the church's need was universal, not merely back in Aarhus, but as much here at the cathedral as elsewhere. He thanked the librarian and went back to the glebe house, guessing at his chances of running into the master of an Irish vessel and perhaps going back with him to Ireland. "It would be closer to Denmark, and once there, I might meet somebody there who could at least carry my manuscript back to Bishop Olsen."

After many one-on-one lessons in Spanish and French, phrases began rolling easily off his tongue. He learned the basics of other languages spoken by the visitors he met, but not the Basque. Their language was unlike anything he could relate to.

When he next encountered the archbishop, Niels listed some details of his progress. "Your Grace, I must say frankly, I am more concerned for the visitors' level of preparation than the language they speak." Shaking his head at the irony, he explained, "These individuals travel long distances and sacrifice much for the opportunity to be here, but they come to The Way so unready. Worse still, I am troubled to discover how the poorest of the crofters are forced to join a pilgrimage to serve a penance in their master's stead."

* * *

Along with the daily hordes of common folk, a rider cantered into the cathedral square with news of Portuguese royalty approaching Santiago. Archbishop Mateo was aware that the counties of Portugal and their various tribes had only recently been united and hoped to strengthen their connection to the church. He put all staff on high alert, insisting the visit of the new queen be their priority. Niels was so excited that his preoccupation with going home faded, and he too became obsessed with the preparations. "We will be in the presence of sizable wealth and a formidable sea power," he told the archbishop, "so learning to speak Portuguese will be my main concern."

When Mathilde, the Queen Consort of Portugal, and her entourage swept into the grand plaza, Niels stood in the first row of greeters, awed by the abundant gilt adorning the royal carriage and by the overwhelming number of servants trailing in an endless caravan of landaus.

Because Niels was so committed to learning Portuguese, including the protocol for greeting royalty, the archbishop selected him to lead the official welcome celebrations. Niels was overcome with nervousness as the stately-coifed lady stepped from her carriage. To his great distress, he bungled his first words of welcome to her in Portuguese. Traumatized, he made an unconscientious switch into Latin, followed by French.

"*Mea culpa … Ma faute, ma grande faute.*"

The queen met his embarrassment with genuine sympathy. "*Mon père*, you speak French. What an unexpected yet refreshing reception. In truth, I was born the Countess of Boulogne in the Kingdom of France. French is my mother tongue." Her eyes widened, and she took his hand. "Please,

Father, would you consider being my personal guide during our stay?"

The queen was enchanted by the magnificent cathedral and followed Niels with great piety during the tour. She was fascinated to hear her young guide had come from Denmark, had learned so many languages, and carried himself with the poise of a much older man.

She fell further under his spell when Niels was granted the rare privilege of leading her into the sepulchre where the bones of the saint himself lay in the sarcophagus supposedly placed there ages before. After a long period of silent reflection with Niels at her side, she explained, "I have a confession to make."

Eager to oblige, Niels responded quickly. "Could there be a more perfect place to make your confession, Your Majesty?"

"No, no," she explained with a self-conscious air. "I am not begging forgiveness for my sins. I have no sins and I don't mean that type of confession, but Father, my life is miserable, not at all what it seems to those on the outside."

Niels couldn't fathom how a lady of such affluence might be at all miserable. "But you are the Queen of Portugal, with servants. Before your marriage, you were a countess in France. You have palaces, numerous ladies-in-waiting and untold wealth."

"I'm not permitted to form bonds with anyone outside the royal court, yet the court has not accepted me. I am not allowed out in public. I am told to hold myself apart. As I am not Portuguese, none of the ladies-in-waiting assigned to me speak my language. I have made few friends and have no confidants in the palace. My marriage to the king was arranged when I was a child and remains the root of the burden I endure. My husband is not concerned with my troubles, so my confession to you is this: I come to you not as a pilgrim, but as a deserted lady hoping to find a companion. I need to express my feelings on, well, personal things. I need a person who is worldly-wise. Someone such as yourself."

She grasped his hand. "Such a learned man as yourself could understand my loneliness? And as a priest, nobody would question you being my companion, an innocent confidant."

Niels was naïve and had no experience with women, but he did suspect her interest in him was not appropriate. He withdrew his hand and tried to change the subject.

"As a priest, I have no expertise to assist you in such matters. The only notable thing I have done is cross what the Norsemen call the Western Ocean, twice. Perhaps I could tell you of my recent journey?"

"What do you mean you crossed the ...? Do you mean what the Spanish mariners call the Northern Ocean? What could be across there but more ocean?"

"Be assured, there is land over there, whatever it is called. A land called Vinland."

"A new land, you say?" Mathilde became enthusiastic and all but forgot her troubles. Niels was buoyed by her interest.

"Not merely a new land, but one surrounded by an endless fishery ...."

"A new land with an endless fishery? Imagine how pleased my court will be with me if I tell them this. My treasury will sponsor an expedition to seek out this wondrous territory. And better still, I will earn the respect of my husband as the first to tell him of it."

"I am so grateful you believe me and will send your fleet to fish the bounty there. I will immediately share all the information I gathered while in Vinland."

"And I am pleased we agree," she asserted. "You will come back with me and describe this opportunity to our fleet masters. Support from the court for our navigators will encourage them to retrace your route and claim these resources for the Kingdom of Portugal."

Niels knew the excellent reputation of the Portuguese navigators. "They are a perfect choice to follow through on this information," he agreed. However, her suggestion of him going with her to Portugal discouraged him. It was in the opposite direction from his home. He tried to decline with grace.

She was not to be rebuffed. "Father, think of the extravagance you will enjoy as my companion. We could continue our pleasant conversations in French, and your information on the fish resource will endear me and you, of course, to my subjects."

He spent the balance of his evening and late into the night agonizing over how to confront her passionate insistence. Hoping a copy of his manuscript would make amends for his reluctance to go with her, he worked through the night preparing a duplicate, including all the facts he could remember of his route, location and the amazing size of the schools. The only missing features were the illuminations. By morning, he had steeled his resolve to remain at the cathedral and present the copy to her.

"If you could deliver this to your court, Your Majesty, your mariners will discover Vinland and treasures in Notre Dame Bay more valuable than any silver mine."

Mathilde was crushed. For a long moment, she stood stony-faced, then stammered, "You are not joining me? Nobody rejects a royal indulgence."

With a merciless sweep of her arm, she summoned her entourage. "I will deposit your manuscript at my royal residence in Coimbra. It is near enough to the seaport of Averio, where some fishermen may take notice." Niels suspected neither the king in Lisbon nor the fleet in Averio would ever hear of it.

"I fear I have now failed a second time. To date, I have not been able to deliver the news of abundant fish resources back to my bishop in Aarhus. Now, I fear any chance of getting the Portuguese fleet involved in the fishery of Vinland has come to naught."

* * *

Archbishop Mateo often observed ladies of royal standing who requested personal tours of the cathedral. He knew young priests sometimes faced demons when squiring ladies of privilege, so he kept a close watch on Niels during this royal visit. He was impressed with all the kindness Niels offered Mathilde, while falling for none of her perilous flirtations.

"Father Niels, you acted quite appropriately with Queen Mathilde. You are a sympathetic listener, and your French diction is excellent. This has given me an idea. I've decided to send you north, where the Basque Country straddles the frontier with France. San Sebastián is several days from here, but the town is where many French pilgrims assemble to begin their walk to our cathedral. They call the route The French Way. There, you will offer them an initial blessing, your advice on preparing for their walk and a souvenir as they depart for home."

Niels could feel his spirits rise. "I appreciate your generosity, Your Grace.

While I have enjoyed assisting you here, I am pleased to hear of other stations where I might serve our church." Privately, he had prayed the day would come when he might encounter a traveler who could take him or at least his writings home.

He had purpose in his steps when he walked back to his cell in the glebe house to collect his few belongings. In his mind, he kept repeating, *I'll be much closer to Aarhus and if the angels are with me, a chance to meet a vessel to carry me home.*

Chapter 16

Tórshavn, Faroe Islands

A flotilla of karves was heading out of the harbor for their fishing grounds when Arne steered the longboat into Tórshavn by the faint light of dawn. Watching the men secure the longboat, Arne had only one warning. "We've been sailing this sea a long time. Now you're on your own until the next moon. You can drink at the alehouse all day and sleep on deck all night — or not. I don't care where you go, as long as you show up onboard when we're ready to leave."

The harbor was shrouded in its usual morning fog, and most of the longboat's crewmen chose to roll their mats out and sleep on the wet deck. Arne also spread his mat on the deck, but under the canvas awning. He hauled a coverlet up to his chin and closed his eyes.

For the next three weeks, the sun rarely broke through. Afternoon downpours that left the townsite soaked could be counted upon to follow the few spells of blue sky. The miserable weather, on top of aimless days drinking and swapping lies with anyone who would listen, demoralized the crew. Arne had complained about the outcome of his last visit to the Faroe Islands, and this one was turning out worse.

"You're right," Stok agreed. "There hasn't been a wench in this place who hasn't rejected me, but one of them did find me a Hnefi board. She showed me how the game is played with tiny 'men' on a square sheet of hide etched with squares." Stok insisted on teaching his brother to play.

* * *

Baldr, too, was on edge, knowing Arne and his longboat crew were underfoot. One evening, as the brothers gathered to drink and bicker over the rules of Hnefi, Baldr marched straight over to Arne and announced, "Pack that up. You can't tell right now, but the moon is cycling back to full. You and I are going out to get your fish and send you back to Ireland." He yanked the rawhide off the table, and thirty-six amber game pieces whizzed in all directions.

"You arse," Stok yelled. "I had Arne's king piece trapped."

"Too bad for you," Baldr snorted.

Stok held one remaining game piece in his fist, and with nowhere left to place it, he flung it at Baldr. The piece flew into a dark corner.

"It's time to go, anyway. I've organized our fishermen to lead you to the Faroe Shoal late tonight. It's an isolated fishing ground to the south where there is a special stock of cod. I'm telling you this because it's a quick way to settle our debt." Baldr stared directly at Arne. "The only thing you have to remember is these fishing grounds are private. We don't tell outsiders. And you can't tell anybody. The merchants' guild in Bergen has never heard of our little fish bank to the south, either. Don't let me hear of you telling them." He jabbed a blunt finger into Arne's chest and pressed until the nail turned bright pink.

Arne hated being touched, but grit his teeth trying not to react. "We don't care where we're going, as long as we get what you agreed to. The fact is, we're tired of this place. Heading south puts us much closer to where we plan to meet up with Aideen's fishermen."

"Aideen? You mean the fishwife in Ireland? Some of our fishermen got in trouble over there and told me there was a lady looking for fresh fish. Go where you want with the fish, we don't care as long as you don't tell anybody where it came from."

* * *

Shortly after midnight, Baldr jumped aboard the longboat and told Arne to follow the convoy of karves out to sea. He signaled when they arrived onsite and within three hours, the Faroese fishermen caught and transferred enough cod to fill the brothers' longboat and filled their own.

Arne was awed. "Filling the boat here is faster than in Vinland, but your fish are smaller than over there."

"Size doesn't matter," Baldr said, and pulled out his half of the tally stick. He passed it to Arne, who matched the two halves and agreed the debt was paid. The two men clasped their lower arms and added traditional, Viking grunts. "We won't be going back to Vinland anymore," Arne said. "There's enough fish here for all the peddlers in Ireland, and this is closer than Vinland."

Baldr was curt. "You will go back to Vinland if you want fish."

Lifting his gaze from the boatload of fish, Arne saw Baldr pulling a knife from the scabbard hidden in his boot. The moon was past full but still bright enough to illuminate the long blade. Balder held it up for a moment, then slipped it back in place. Its threat was clear.

Masking his irritation, Arne asked with a hint of sarcasm, "Was that a dagger I saw? You still have the spirit of a Viking, you old troll." Baldr jumped over to his own boat and repeated his warning. "If you are ever seen chewing on a fish from this bank, it'll be the fish feeding on you next. We only brought you here tonight because of all the farm tools you left with us. The guild knows nothing of it, and they better not hear anything of this arrangement either."

As the fleet of karves moved off, Arne relaxed. He had the fresh fish he could sell in

Dingle, and now he knew Baldr's secret — an advantage he would put to use later.

* * *

Stok called over to his brother. "I'm liking this. It's the easiest we have filled our boat yet. And what else do I see? Are you smiling?"

Arne's grin evaporated, replaced by a concerned stare. "We collected what we came for. Now, if these winds are right, we'll meet Aideen's fishing boats late tomorrow and collect what she promised us."

"I'll be happy to see that fair maid again." Hoping to emphasize his comment, Stok clomped his wooden leg on the deck repeatedly until the rounded end of his peg stabbed the slippery head of a cod. He lost his balance and fell in a heap.

Watching Stok flounder in the mass of fish, Arne's smile returned.

"There. That happened because all you do is think of women."

Stok was scraping fish scales from the mermaids carved into his peg leg. He struggled to regain his footing and kept picking guts off his tunic.

"Once I'm all cleaned up, there'll be a wench waiting for me in Dingle with a tall pitcher full of ale, maybe a bottomless one. You can count on that, brother."

"You won't drink and I won't sleep until we see those boats outside Dingle Harbor.

The ones Aideen assured us would be waiting. Put the men to work cleaning and splitting this load."

* * *

By the following midnight, the longboat was two leagues off the Irish coast. Arne could make out the opening to Dingle Bay between the peaks of two low mountain ranges. Standing ankle deep in the mass of cleaned cod, fish scales glittered in the faint light, reminding him of how the bishop's ring sparkled at the launching ceremony in Aarhus.

Emerging from the gloom, five small vessels approached the longboat, each carrying full barrels of salt and empty tubs for the fish. As they drew close, Sean shouted, "Even in this faint light, I hope you see now how a little honest work rewards all involved."

Arne shot back, "What I see is you're surprised we did what we said we would."

With all the vessels rafted up, the load of Faroese cod was forked over to the open currachs. Sean's fishermen marveled at the quantity as they fit them into the tubs to begin layering on the salt. Arne craned his neck over the side to watch, "Tell me again what your salt does for the fish."

"With the right amount, the moisture is drawn out and the fillet dries quicker. Fish with salt keep for months, maybe longer.

Watch." He sprinkled a half shovel full of the salt. "You put a full measure there where the fish is fattest and less down at the tail."

"What's the matter with fish hung to dry in the air? We call it stockfish and carry it aboard wherever. It won't go bad in months, or ever. It's always ready to eat."

"We call it Viking fish," Sean admitted. "Nobody here eats it anymore because after you hack pieces into thin strips with an ax and soak them in mother's milk, you might still crack a tooth on it." Speaking loud enough for his fellow fishermen to hear, he added, "I'd rather suck on my own fat thumb." His fishermen all wanted to laugh but waited to see if Arne was offended. The big Viking ignored the comment.

"And anyway, we enjoy the taste of the fish when it comes out of the pickle," Sean added.

Arne shrugged. He hadn't considered that. "But it's a lot of extra work."

"What else are you going to do while you're sailing back into port? You could be adding value to your product."

"But you have to pay for the salt."

"No, we don't. Aideen gives us the salt, plus these empty tubs. We fill them, spread the salt she gives us, and she sells most of her finished product over to Bessie's or to locals who walk right out there to her place."

Stok butted in. "Where's the salt come from?"

"From the sea," answered Sean.

Stok frowned. "Yeah, yeah, but how do you get the salt out of the sea?"

"Aideen brings in loads from up the coast. We get a lot of high storm surges this time of year, and the folks on the Aran Islands scrape it out of the sea pools between the rocks after the spray has evaporated. A stormy spot those Aran Islands are. She pays for it with the vicar's beans or whatever she's able to barter. Aideen's always bartering. With her, it's one thing in exchange for another. We're all links in her chain."

Arne stared directly at Sean. "What if we got the salt ourselves, went back to pick up the free fish in the Faroes and took it right into Limerick, sprinkling salt on the way?"

Sean balked. "Hey. I've said too much. You can see how this is a valuable business, but it's her business. If you two plan to do the same, why not think of doing it in your own country? I bet if you took this salted product back to Norway, they wouldn't want your Viking fish anymore."

Chapter 17

Dingle, Ireland

Long brown fronds of kelp wrapped around the pillars of Aideen's wharf. In her recurring dream, they magically transformed into a whirl of codfish until she reached deep into the cold water to grab one. She sat upright, fully awake and wondering, *does this mean all the fish I've waited for will appear, but I will never land one?* She rolled over and buried her head in her bolster. *How am I going to fill my warehouse if I can't even hold onto a fish in my dreams?*

The longboat bumped against the far end of her dock. She looked up from the bolster with a start, spying a beautiful bird's egg blue sky outside the garret windows. It was a view Aideen always enjoyed on the fogless mornings along her shore, but the peace of the morning view was short-lived. Arne bellowed from the other end of the dock.

"I want this deck to smell fresher than new oak planks. There's sure to be an alehouse somewhere in this village, but there'll be nothing to drink for any of you until this boat is scrubbed clean of fish guts."

Aideen jumped up, grabbed her shawl, pulled it tight around her shoulders and stepped out. "Welcome to Dingle," she called, trying to sound as if she expected to be seeing them that very morning. "This is the same time of the day as our meeting last month at Limerick's nunnery. At least there is no abbess here to be upset by the ruckus. Come inside Arne."

Arne followed her into her near-empty storeroom. He batted his arms at the barn swallows darting over his head, parked himself on an empty keg and sniffed the air.

"Is that your salted cod I smell?"

"Yes Arne. The little of it the fishermen here have been able to bring in to me."

Arne shivered. "It's summer outside. Why is it cold in here?"

"The sea breeze coming through the louvers is damp, and there are no windows to let light in, so the salt fish cool until we put it out to dry. Hopefully, Sean and the other boats are on their way in with your fish, and this room will be full by tonight. The stronger the smell, the greater my reward."

"My reward will be a keg similar to this one I'm sitting on, full of coins."

Aideen lifted a weight off a stack of fish and took a fillet out of the brine. "Here, try this. I've been saving it for you to taste."

Arne pulled out his knife to carve a strip off.

"You'll hardly need a blade. Shred a corner off with your fingers."

Arne studied it, leaned in with more interest and ripped a handful off the fish flake. With his mouth full, he asked, "How long would this keep at sea?"

"I have no idea. It doesn't stay around long enough to find out."

"Your fisherman, Sean ... he says you do well with this business, right?"

"If you keep delivering it to me, we could *both* be doing well." She lobbed a hand-knit sock full of copper tokens into his lap. "Here's a first payment for your crew."

Arne stopped chewing and sifted his fingers through the contents of the sock. A vacant expression covered his face. Aideen waited for a more enthusiastic reaction. It didn't come.

"I can't tell what you're thinking, but here's what I'm suggesting. When we are done here, you and your men hike down the strand to Bessie's. There are enough coppers in there to trade for all the drink I know they'll want. While you're down there enjoying her brew, I'll gather up the salt you need for your next trip to the Faroe Islands. You'll be ready again to do our deal next month, right? It'll be the harvest moon — a bright night."

Still, Arne offered no reaction. Aideen pulled a full basket of fish over. "This is for your crew, too. Take it with you to Bessie's and have them all try it."

Arne rapped his knuckles on the empty keg underneath him. His thick, blackened nails made a hollow ticking sound.

"Ah." Aideen nodded her head. "Evidently, you're thinking copper tokens are hardly payment for your part of the deal and you're right. There'll be a fair share of silver waiting for you after Sean and his fishermen come in with the load you transferred to them. First, let's see how many kegs they bring me before we settle up."

Arne stood and walked back to the longboat with the sock of coppers and the basket of fish. "Your man Sean and his boats are somewhere behind us," he shouted over his shoulder. "There are five of them, all full of the sweetest, slipperiest cod you ever saw. To be clear now, along with ale, silver is my favorite thing, and I won't be leaving Dingle without it." He climbed up onto the longboat and called Stok over.

"Your pretty Irish fish wench in there is trying to teach us how to buy and sell. I'll explain later down at the brewhouse. It's probably the only place we can spend these coppers."

He shouted to the men swabbing the deck with wet rags and buckets, "Once you're done, you can all join us for a meal."

Aideen stepped out and waved an arm in the direction of the inner harbor. "The brewhouse is on the strand. I'll walk partway to show you where."

* * *

Aideen set off down the shore with Arne and Stok, but had only gone a short distance when Poppie called her back. "Sean and the fleet have arrived. They are tying up now and waiting carefully for you. You should see their full loads all packed into kegs and topped up with salt."

Up on the wharf, Poppie operated the davit arm and hoist, uploading each of the kegs from the small boats and directing where to stack them inside. Aideen was amazed how quickly the wall disappeared from sight, along with half the floorspace now covered with row upon row of kegs, two tiers high. It was more fish than she had expected.

"Poppie, I remember how you gave us some bandy-legged steps across this floor a few months ago. In many ways, I'm glad there's no room for dancing now."

"If it's another dance you're asking me for, Aideen, I could move a few of these kegs."

"No, no Poppie. You've done a great job getting all these fish stacked in here for me, but there'll be no room for a *céilí* to celebrate our success. Bessie wouldn't be available to play her whistle, anyway. She's too busy dealing with the crew of thirsty deckhands off Arne's longboat. Remember how worried I was we'd have nothing to fill this big space."

* * *

Arne and Stok marched into Bessie's and informed her that their whole crew would soon be on the way in. "Better get some more wenches in and draw off as much ale as you have," advised Stok.

Stok was disappointed when Bessie's man, not a barmaid, appeared carrying the full tray of earthenware flagons and two jars to the table, but he did notice Bessie working at the kettle in the back. He downed his first full jar and strolled over to flirt with her, only to be told she understood no Norse. He persisted until Arne flung a copper disk at him. It bounced off his ear. Stok winced, and Bessie tried to hide a smile. The few locals in the place appreciated the spectacle.

Stok cursed, rubbing his ear until he noticed Bessie was enjoying his pain. He gave her his best toothy smile, but it made no impression.

"Leave the wench be, Stok. Come back and sit down. We have to talk." Arne slid the full sock of tokens across the table. "Here. Give her this when the crew comes in. They can drink 'til these coppers are gone. You too, but you'll need a lot more than copper if you're going to turn her eye."

Stok was still rubbing his ear. "If I had a chance to learn the local language, I'd show you how to arouse the busiest of bar wenches."

Arne leaned in and whispered, "Aideen says she has silver for us and will give us free salt after she tallies the load her fishermen are bringing in. And you know what she asked me? 'Will you be ready to go out again next month?' I didn't tell her yes, and I didn't tell her no."

"So what do you mean, Arne? We don't plan to keep running fish from the Faroe Islands over here to Aideen? It's easy work."

"We agreed to bring her this load because we made a deal with Baldr for free fish. I'm saying we'll have her silver and a load of free salt, but there won't be another load for Aideen unless we have another trip of free fish."

Stok stopped rubbing his ear. "We know Baldr won't go for that again. What are we going to do with all the salt you say she's giving us?"

"Remember how the fishermen told us they always salt their fish on the way in?"

"What? When?"

"Listen Stok. Right now Aideen is loading her salt aboard our boat. What if we go to Vinland with it and get a free load of the bigger fish? We can layer on the salt ourselves the way Sean showed us. Afterwards, all we'll have to do is ride the river up to Limerick and sell it to the same peddler. He'll pay us more this time, and who's to say where we've been or where we're going with it."

* * *

Benji and Dag ambled into Bessie's at the same moment as Arne mentioned Vinland.

"Great news for me, Arne." Benji clicked his heels on the floor. "We'll be going back to Greenland on the way."

Arne sputtered. "I didn't say Greenland. I said Vinland where the fishing is easy. The fish are much bigger than the Faroese cod and it's not too far if we don't go out of our way to Greenland." He lowered his voice. "We'll do the catching over there and the selling back in Limerick. But remember, we're telling nobody where those fish are and where we're selling them."

The crew kept filing in, crowding together on the available benches. Olaf was carrying Aideen's basket of fish and distributed it while they waited for Bessie's husband to deliver more flagons of ale.

Bessie carried another tray full of ale over to Stok and waited for him to pay for all the ale the crew ordered. "I guess you'll want the rest of these coppers," grumbled Stok. He handed the sock to her with a wink. She thanked him in perfect Norse.

Bessie's quick response startled Stok. He turned to his brother, grinning. "See. She wants to talk to me after all."

Arne scoffed, "She's a bar wench. She learned to handle a souse like you long ago and, in any man's language."

***

The balance of the day was spent swishing down Aideen's fish with Bessie's brew. To a man, the crew bragged about the way the long trip had worked out for them, except Dag. He slid in beside Arne, whose cheeks bulged from chewing the Fish Balls Bessie had plunked in front of him moments before.

"Arne, I see you're enjoying this salt fish. We all are. Now we know how good it tastes, shouldn't we take the next load to Norway? We'd have no problem selling it in Bergen. The guild would buy it at a higher price than Aideen gives you."

"Norwegians want dried fish ... stockfish," Arne barked. "You keep this between us, but we're going to sell our next load of salted fish to the peddler in Limerick. The plan is the plan."

"But if we make our way to Bergen and show the buyers this product," Dag persisted, "they'd want you to change your plan."

Arne was half convinced Dag's idea might be better than his, but he resented not having come up with it himself. "We're heading home to Stavanger for the winter. We'll drop off our recent earnings at the island hideout and save a fat reward for the two men we abandoned with the priest off the coast of Ireland. Why go up to Bergen? It's another day north."

"Because Bergen is the center of the fishing world. The guild in Bergen has a royal charter, a monopoly on all fish throughout the Norse lands. I mean every fish, from water to table. Baldr tried to tell you, and I'm from Bergen, so I'll tell you too. I've seen how they run things," Dag explained. "Every fish goes through their warehouses, and the profit stays right there. But if they were to find out how good this is, they would pay more to us than they pay for the dried stockfish, probably more than the treasures you have stashed away in the hideout. You know, they'd probably make you a partner."

Arne was blunt. "We've been all the way to Vinland and back. Now we're going home."

"You're always saying you are at home on the sea."

"What I'm always saying is I like getting things for *free*. If those merchants have a monopoly, they'll demand a percentage, and I'm not paying them any fees. They can wait until their own fishermen come in." He picked up a flake. "Dag, look at this. We can go anywhere we want in Norway to sell this salted fish."

"No. Their guild is too powerful. They'll discover what we're doing and shut down any buyers who deal with us."

The idea of a handful of men powerful enough to control all the Norse lands impressed Arne. He took another bite and waved the rest under Dag's nose. "Here's what we'll do. I won't be paying a fee to anybody to sell my own fish, but maybe we can spare an extra day to go to Bergen to see what they offer us. If they decide they want what we'll have for their market next spring, they should pay us. In fact, there'll be an extra charge because we are the only ones who know where to go."

* * *

Vicar Maurice's spy reported how Sean and his fleet had landed a large volume of valuable codfish at Aideen's. "It's not only Sean and the local fishermen, either," explained the informant. "A longboat docked at daybreak to load on supplies, and once they finished, the rowdies headed over to Bessie's, overwhelming her place. They're still there, swigging her ale and filling her purse."

Maurice groused to his cronies, "As Village Magistrate, I should be getting a part of all this profit making, but I'm not. Instead, my purse becomes ever more thin."

Maurice couldn't begrudge the success of any business in his parish, but privately he resented hearing Aideen and Bessie hailed as the new leaders of County Kerry. "And they're both women," he groused.

He called for his palfrey to be brought around and prepared for a trip down to Beenbawn so he could confront Aideen straight away.

Still well back of the storehouse, Maurice was shocked by the number of kegs he could see being wheeled around the site. He spotted Aideen trying to keep a tally of the incoming kegs. Remaining in the saddle, he huffed loudly. "I knew you'd be no ordinary fishwife. Is there no end to the fine living you are now making? Since you are doing so well, but yet paying no taxes, I'll review the parish records to ensure you are paying increased tithes on all this."

Maurice squirmed on the lambskin his page had placed on his saddle and exhaled heavily. Aideen caught a whiff of rosewater and knew he must have been bathing in the local concoction purported to relieve the discomfort of hemorrhoids. She kept silent.

"And since my treasury is in shortfall, I am increasing your rent. You have already paid me the shilling for this year, but beginning next year, you will pay me five shillings *per annum*."

"You don't know my business, and you can't pick a number out of the air."

Maurice gestured at the activity around them. "Considering all this, it's more than fair. I understand your rent was set at one shilling because you earned scarcely more than a pittance from this bogland. Nobody could have predicted what your calculating mind would turn this landholding into. But now, I estimate you are making much more than fivefold the earnings you ever did on bricks of peat; therefore, the rent will be five times the —."

"And where do my earnings go, Vicar? Sure you've had your deacons check through the parish records. They go to these fisherfolk who receive a decent price for the first time. From there, they go into the hands of the innkeeper, who keeps my fisherfolk and her customers well-served. Thanks to that, the stewards of your congregation collect greater tithes from us all. It's a rare pence that multiplies itself for the benefit of so many, Vicar. Am I not right?"

"A rare pence indeed." The faintest smirk appeared at the corner of his mouth. He yanked on the reins. "You'll pay me five shillings *per annum,* more than fair, in my opinion, as I'm still not accounting for my mule and cart behind us there."

Regardless of how logical her argument was, Aideen knew the outcome would be the same.

* * *

After the crew's boisterous evening, Arne was eager to leave at daybreak. He told the crew to ready the boat for sea, banged on the outside walls of the storehouse to rouse Aideen from her sleep and demanded they climb into her rafters and count out his final payment in pieces of silver.

While Arne was inside with Aideen, Dag explained Arne's new scheme to Benji and how the brothers planned to slip away with Aideen's salt, never to return. Benji saw little chance to warn Aideen and a real possibility that Arne would be infuriated if he found out, but he had to try. He slipped into the warehouse and waited under the staircase, listening to the conversation overhead.

Aideen handed over a pouch of coins. Arne felt its weight, opened the drawstrings and sifted through the contents, listening for the familiar chatter made by the low-grade Norse silver. Satisfied with the payment, Arne grabbed another basket of salted fish on the way out and climbed back aboard the longboat.

Benji felt uneasy approaching a grown woman, something he had never done growing up in the tiny Eastern Settlement, but rehearsed what he planned to say as he climbed up to face Aideen in the rafters.

"Ah, Aideen. You, ah, might remember me. We met at Tilly's Alehouse in Limerick. You asked me for fresh fish, and I was too busy talking to pay you any mind. I don't recall now what you asked us, but I do remember I was too nervous to thank you for the ale you bought."

Aideen smiled. "Not to worry, my tall blond friend. The truth is, when we met in Limerick, besides being on the hunt for codfish, I wanted to have an up-close look at you. The good news is, the next morning, I was to make this deal with Arne. As a result, my storehouse is full. If it's the same next month, I'll need to bust out a wall."

"That's the trouble. Next month it won't be the same. We're not going back to the Faroes. Arne didn't tell you, but he's keeping your salt and doesn't plan on bringing you any more fish. You'll not be seeing us again. We're going to Norway for the winter, and next spring we'll be going back to Vinland. Then we'll salt and sell the next load in Limerick."

Benji could see his words had struck her like a gust of icy air. This time, he held her gaze and tried to be positive. "I wanted you to know, somehow I'll do something to make things right."

Aideen took one of his hands in both of hers and squeezed gently. "Thanks for the warning, but don't worry. I don't make deals I can't afford to lose," she lied. "And I have wine coming soon from the Spaniards to make up for it." Another lie.

Benji could still feel her soft hands gripping his and was sure he could hear the thrum pounding in his chest as he hurried back to the longboat. For the first time in his life, he had overcome his fear of talking with a woman. His worry now was whether Arne might suspect he had been inside the storehouse, telling tales.

Still thinking of how he might someday make it up to her, he bumped right into Olaf and saw a suspicious leer spread across the big Viking's face.

"I had to get rid of all the ale I drank last night," Benji lied. He stepped over to his bench in line with the other rowers and helped swing the longboat out to sea.

Chapter 18

San Sebastián, Spain
The city of San Sebastián was known for centuries as the busiest fishing port between Spanish Castile and the various French kingdoms. But after the autumn harvest, the many markets in the center of town didn't smell of the usual fresh seafood. They were overwhelmed by the aromas of fresh-cut flowers, fruits, vegetables and seafood tapas with cheese and cider. At the same time, the waterfront burst with pilgrims, mostly French subjects, plus a few caravans of Anglo-Normans and an occasional noble with his servants in tow. They gathered to begin the autumn trek to Santiago de Compostela, following the route called The French Way.

Shortly after daybreak, Niels attended to his personal liturgy and walked to the harbor. As he walked, he hoped the same hope. *Perhaps this is the day I encounter the captain of a caravel who will offer me passage home. Good outcomes are only a matter of earnest prayer, right?*

A foreign caravel arriving with pilgrims was rare, Niels learned, because most pilgrims came into town on foot. To greet them, the tall priest donned his black robes and hurried to the marketplaces, offering his initial speech in Latin, followed by another in their home language.

"Greetings, my children," he would begin. "I am Father Niels, and I welcome you to a grand awakening." He suspected the travelers wanted only a short blessing before hiking the long and poorly-marked path, but he had to take extra time to break the harsh truth to them. "You have made some sacrifices to come this far, but I doubt you prepared yourselves for weeks of walking across the rough terrain and into the province of Galicia. I must caution you, The French Way may be shorter than the traditional Way of St. James, but it is equally as arduous. Painless pilgrimages don't exist — blistered feet do. Be assured, though, the memories you make will soothe the difficulties of your final years."

"There is also a gift," he added with sincerity. "You will receive an inscribed scallop shell signifying our promise that your required interval in Purgatory will be shortened, whenever that should arrive for you." He concluded with another incantation in Latin and wished the new arrivals *Via Con Dios*.

He had prepared an embellished version in Norse, for the slim chance any of his own countrymen might appear to hear it. To date, there had been no such visitors.

* * *

For the fishing fleets at the port, autumn marked the end of the herring season and weeks of preparation before their extended whaling trips. After storing their fishing gear and preparing the equipment for whaling, Basque fishermen often gathered on the pier to watch the arriving pilgrims. Captain Gabriel de Portu was among them. Widowed and with time to spare, he often sat, parked on a coil of rope, taking in the priest's presentation. His ship's mate toiled beside him, most times knotting a monkey fist on the end of a new hemp line. Because his mate was shorter than most men, de Portu had given him the affectionate nickname 'Shicki'. The man accepted that cheerfully, and was now known as Shicki to everyone.

"Shicki, where do you suppose the priest picked up such a peculiar accent for his talks to the crowd? With each speech, his voice gets louder and his accent grows worse."

Shicki twirled the completed monkey's fist over his head. "Who knows, Captain? Why don't you ask him?"

"You're right my friend. I should." De Portu stood, and the two walked over to where the priest was finishing his blessing of the latest crowd of newcomers.

"Hello Father. I am Captain de Portu and this is Shicki, my mate on the whaling caravel *Espiritu*. We are moored up at the quay there. We see many priests in our town, but I must admit, it is rare to see anyone who mangles foreign languages quite as you do. You make a sincere effort, but nobody understands a thing you're saying. Where is it you hail from?"

"My archbishop in Santiago sent me here hoping to put these new arrivals at ease. With my personal experience at sea, I was assigned to welcome these pilgrims who arrive by boat."

"But you are not *from* Santiago. You have a Norse accent." De Portu switched into his fractured Norse. "Where were you born?"

Niels brightened at the sound of his native tongue. "So. You, too, can easily mangle another man's language."

Shicki was amused by the priest's impolite retort. De Portu frowned.

Niels knitted his fingers and with a pious bow said, "Forgive me. I couldn't resist the quick barb at your expense. In truth, I was a simple parish priest assigned to a growing flock in Aarhus, Denmark, but I've since been clear across the ocean and back after I left there this spring. Perhaps my accent reflects all the places I have been and the many different people I have spoken with since."

"Ah, but Father, you're a long way from Denmark. How did you end up here? Maybe I should ask first, how did you cross the ocean? More important perhaps, why?"

***

Being with experienced men of the sea, Niels felt he could relate his story to these men, who would not only understand his trials but also be compassionate when learning of the most harrowing aspects.

"At first, my mission was a dutiful one. You see, my bishop arranged for me to join a fishing expedition heading west. It took many weeks, rowing most of the way to Greenland, then on to Vinland. For most of the time I was laid low with seasickness. At long last my reward was in coming face-to-face with the Forest People. Sad to say, our captain got his boatload of oversized fish that very afternoon and insisted we return right away. I had no chance to make a genuine connection with the Forest People."

"Wait! Your story is scarcely believable, but the part about a boatload of fish. How fast did your captain fill his boat?"

Niels was unfazed. "Our vessel was in Notre Dame Bay for one afternoon, maybe less because our captain spent half the day arguing with some of the Greenland fishermen."

"The fish must have been jumping into your boat," De Portu interrupted. "What could your captain argue over?"

"I paid little attention, but isn't one afternoon enough to hook a boatload of fish?"

"Father, most of us Basques spend weeks or more at sea trying to hook enough fish to *eat*. Think of all the days we waste hunting for a single whale and then many more days on land cutting up the meat and rendering the blubber into oil before we are paid. If I could fill this boat in one day, I'd wait no longer and set off today. Tell me, where might I find such bounty?"

"First, you must understand I am a man of God, not of the sea. My whole voyage was in service of my faith. I was required to gather information on the fishing activity in the hope we could feed our congregations during Lent. I did what I could, although I cannot be certain the fishermen told me the complete truth. For some reason, they wanted to keep the location secret. Sadly, it will remain so because on the return trip, we suffered a shipwreck and all my records fell into the water."

De Portu sighed. "But you are here now. You must have been rescued. Try to remember something. First, how did you get to this Notre Dame Bay? The location, Father, is the key."

Niels became exasperated by de Portu's questioning and struggled to think what he wanted to say next. "If we hadn't been sinking, if I hadn't lost all my valuables ...." He exhaled, crossed himself and started over.

"Our location, as best I remember, was sixty-three degrees north latitude. That's what was written in the manuscript I deposited within the library at the cathedral. Regrettably for most who will see it, the text is in Latin, but I asked for a few illuminations in the margins."

De Portu sounded discouraged. "After years going to mass, I now avoid Latin. It reminds me of church. What would encourage you to return to your library and make a translation for me?"

Niels lit up. "You could take me home on this whaling vessel of yours."

De Portu was blunt. "If I receive information on the location of as many fish as you say, 'a boatload in less than one day', I'll not go out of my way to Denmark, I'll hustle straight to those grounds."

Niels accepted the disappointing response, shrugged and shuffled back to the wharf.

* * *

De Portu once heard mention of an astonishing number of fish in old Viking sagas. It all sounded similar to what this Danish priest was telling him now. He called Niels back.

"Father, no real fisherman could let this chance go by. You hold the key to the quest of the ages. Call it a gift from Heaven. We could fill our boat near Greenland and then sail back to a buyer I know in Ireland."

An image of the lady fishmonger standing on her half-built dock in Dingle flashed into de Portu's mind. He grabbed Niels' arm and shouted, "Ireland is better than half the distance to your home. I'll offer to carry you at least to my Irish fishwife's place of business. There would be another ride on to Denmark from Ireland, yes?"

Niels often thought about how Arne had assured him that any fishermen who rescued him would take him straight to Ireland. He was tempted to accept de Portu's offer on the spot, but still he felt obliged to fulfill his promise to the archbishop. "As much as I have prayed for such a chance, how could I abandon my duties for a trip of only half the way to my home?"

"Perhaps we can offer you a little more. If you go back to your library and copy the details of the fishery you included in your manuscript, we will offer you passage to Ireland. I'll talk to somebody I know there, an importer of wines. She'll organize the remainder of your passage to Denmark."

De Portu studied the priest but detected no sign of real interest. Hoping to convince him to at least share the location of Notre Dame Bay, he stretched the truth. "This importer I know in Ireland, she has close connections with the traders' caravans connecting to England and beyond. She would arrange a speedy passage home for you."

Niels was unmoved. "My archbishop back in Santiago assigned me this duty, and I must attend to the many arrivals here. You see how I am able to assist them."

"You can believe me when I say the connection I have in Ireland is a God-fearing woman with similar goals as you. She is making a great effort to import both fish and wine for her brethren in the community. If you were to accompany us, you could bless the wine we carry for her and sanctify it for holy communion, all before you continue on to your appointed journey home."

After a moment, the young priest understood where de Portu was leading him. "Yes, Captain. I grant you the vineyards of Spain are a blessing, and I admit they have become a personal indulgence of mine; however, I cannot go back to the cathedral on my own initiative. My supervisor must recall me first. When he does, I will spend a night at the library and copy the specific information you request."

De Portu exhaled with relief. He could afford to wait a few months, at least until the start of the fishing season, but needed to cover his first lie with another.

"My contact there has warned us that there are import taxes on both wine and fish in Ireland. She hopes to avoid paying them both, so you must say nothing about the true nature of our cargo."

"As I told you, Captain, wine is a blessing and should not be taxed."

***

Weeks later, Niels walked up to the wharf and handed de Portu a scroll of vellum marked with notes and a map indicating locations, plus the number of travel days. De Portu quickly retreated to his cabin to form a plan. Bending over the crude map, he worked out how *Espiritu* might cross what the priest marked as the 'Western Ocean.'

"Shicki. Come and read this information the priest fetched for us. This is a route to riches. We can now locate the schools of fish mentioned in the sagas."

"Yes Captain, the map does show how the priest island-hopped to Iceland and Greenland, then headed south to arrive at a bay he named for Notre Dame. But see there, his manuscript says he took eight weeks to do it, and that was only *one way*."

De Portu wasn't listening. "Think of this. Going over, we'll ignore those islands he visited. They are too far north to bother with, but read the priest's notes in the margins. Once in Greenland, Vinland was a matter of three days south. Now here's what I'm thinking." The captain unrolled his own chart of the known coast and drew a line outwards along sixty-three degrees north latitude. "I believe the priest must have been lied to by the Vikings because his chart showed the latitude of Vinland to be sixty-three degrees."

De Portu laid Niels' notes on top of the charts. "We know Iceland is sixty-three degrees latitude, and Greenland is below that. The priest went farther south to Vinland, so how could Vinland be sixty-three degrees? With a little understanding of the night sky, you and I can figure out its real position. I'll wager the location to be closer to fifty-two degrees north latitude.

"If we set off for Ireland as usual, turn west when we get near and follow a straight heading, say ... fifty-two degrees, we won't need to go to Greenland first, saving us hundreds of leagues and at least a week of sea time." The captain returned his finger across his own chart, showing Ireland to be at the same fifty-two degrees of latitude.

Shicki was not convinced. "And you're predicting all this from the recollections of a seasick priest?"

"You're right, Shicki. This priest is no man of the sea, but those Vikings aren't the fine sailors they think they are either. That's the reason they'll never take this fishery seriously. They took eight weeks to make the crossing because their longboats aren't made for heading into the wind. Look at the priest's notes, he says they rowed most of the way."

Shicki was curious, but still not persuaded. "Do you think our vessel could cover the distance much faster than the Vikings with all their rowers? And remember, it's often into the wind."

De Portu was adamant. "You watch. We'll do it in half the time. And we must do so before the priest talks to anybody else. His manuscript is dedicated to the Queen of Portugal, and if the Portuguese see this information, no matter how sketchy or wrong parts of it may be, it's still enough to light a fire under her fishermen. They have a head for navigation. Vinland will become a national priority for them. They'll discover this Notre Dame Bay and hook the last fish before we get there."

As close friends, Shicki knew the threat of Portuguese fishermen getting a step ahead of de Portu would awaken his competitive temperament. "Captain, how could we sail all the way to Notre Dame Bay and return with fish still fresh? Will we make this vessel of yours soar farther and faster than a seagull?" Shicki asked. "And say we could fly; three weeks is still too long to keep fish on board. That's why the Norsemen dried it hard for their long voyages. Do you want that? Nobody would buy it."

De Portu wagged his finger. "Are you forgetting the redheaded fish buyer we met in Dingle? She told us 'salting is the key to carrying fish'. It keeps better. What if we partly dry and salt the fish we catch? It will cure on the way homeward."

"But Captain, salted cod is hardly in demand here. We won't be able to sell it. We Basques have more than enough fresh seafood. Think of the mackerel and sardines and all the scallops. Yes, we cure our pork with salt, and maybe we eat salted whale meat with peas, but Captain, salt fish is unknown here. It's *fresh* fish you'll see on the table at my house, lots of it. We don't eat cod at all, much less after it's salted."

"I am aware of that Shicki. Haven't I already suggested we need a new quest? A trip for a boatload of salted cod is new."

"And who controls the source of all the salt? The high office of the Pope in Rome is who. He'll not let it go cheaply to poor whalers."

"We're not going to be poor whalers any longer. Aideen, our fishmonger friend in Ireland, is waiting for us to give us all the salt we need, and she'll buy most of our catch for a good price."

"So, are we going for the fish or are you wanting to court your Irish lady friend?"

"I admit I was smitten. In my defense, I have not taken well to being a widower this past year. Still, you cannot ignore the fact of this map. It is a route to riches."

"Even so," Shicki responded, "your crew has no interest in going to Ireland, let alone the other side of an uncharted expanse such as the priest describes. Nobody dares go much beyond the sight of land. We'll have to convince the men our good ship *Espiritu* will be so heavy with fish on our return, they'll need to pound in extra rivets to keep the planks from splitting away from the boat's ribs."

"We'll deal with the crew," de Portu confirmed, "but not yet."

* * *

The popular *Bar La Navarra* was purposely located directly across from the busy fishermen's pier in San Sebastián. It was open for business to serve the dock workers whenever boats filled the port and thirsty fishermen almost anytime. Particularly at the setting of the sun each evening, the bar would fill. It was the 'Cider Moment' — a tradition in San Sabastián when the captain must buy his crew the first round of cider. Respecting the tradition, de Portu arrived to join his crewmen this evening, to buy the cider and announce what he had planned for the coming winter fishery.

"As usual, we'll cruise along this coast," de Portu explained. "I've received a fair offer for any whale oil we might collect, and you'll all share as before. If we find but one of the big black beasts, it'll still make a fair season for us."

The captain checked the faces around the table. Each member of his crew accepted the news, except Domingo Elkano. He was a seasoned deckhand and the Second Mate on *Espiritu*. He often acted as the crew's spokesman, and de Portu expected a provocative comment to come his way.

"Captain, what if we happen to secure a second whale? I guess you'll have to order twice the cider upon our return," teased Elkano. A cheer echoed across the room.

De Portu was pleased with their enthusiasm and took the chance to hint at what changes he had in mind for the following summer. His tone became serious, "Mind you, we'll need to stop hunting whales earlier than usual to take *Espiritu* over to Hendaye." Hendaye was a humble hamlet on the French side of the Basquelands where de Portu had been born.

"She won't need a complete overhaul, but this will be a longer trip than usual and we want our caravel's holds ready to be filled right up next summer." Another cheer filled the room. Nobody picked up on de Portu's use of the words 'a longer trip'. Even Elkano didn't suspect their captain was preparing the vessel to head across the uncharted ocean.

As usual, the discussion around the table turned to what they would do with their future earnings. As the men talked, de Portu pulled Shicki aside.

"There is one element to this plan I haven't mentioned yet. We'll visit Hendaye in the spring. It's the best yard in all of France for a refit, but more important to me is to reunite with an old sailmaker there who possesses a rare skill. If we are going to 'soar farther and faster than a seagull,' we'll need a new mainsail to make a real flight possible."

"A word of advice from me, Captain. You'd better tell Elkano and get him to agree first. You know how the crew always follows his lead. If he says no, how many others will be willing to leave the sight of land so far behind?"

More cider was ordered along with extra servings of fish stew. As they ate, de Portu whispered to Shicki, "You and I can deal with Elkano next spring, but there is something I should do now." He stood up and went over to chat with Sancho, the bar owner.

"Tell the kitchen to spread extra salt into the fish stew. It'll get them used to the taste."

* * *

After leaving the cider house, the captain took an overnight ride up the coast to Hendaye with a wine maker who was hoping to buy aged oaken casks from the traditional coopers in France. "It's for this coming year's harvest," the man explained. "I put my fresh Spanish wine in old French casks, and after it ages, the wine is truly Basque ... three parts Spanish and two parts French."

Early the next morning, de Portu stepped ashore in the familiar French seaport and climbed alone up to the legendary sail loft. The stair treads creaked underneath his steps. Heady scents of flax fiber, hemp and oakum filled the air.

At the top of the stairs, a true artisan with a full head of white hair caught sight of him through the open door of his loft. The man grinned broadly at the sight of his favorite client. "Gabriel. It has been too long since I saw you, and you come home so rarely these days. Come in." He slipped off his sailmaker's palm pad and grasped de Portu's hand.

"Shoo, you old hound," he said to his dog, clearing a place on the settee for de Portu to sit.

"Have you chased too many whales already? Is the canvas I sewed for you last year all frayed and torn?"

"The main sheet you made for us is still fine. My crew admits if it didn't trap the wind so well, they would have to row like Vikings. I do need a new sail for next year though, but not one to *trap* the wind. This year I will need one to speed my good vessel straight into the wind. I hear the Latins on the Mediterranean had such designs. I need you to make me one over the winter?"

"You mean a lateen cut and yes, the Moors invented it. They cut their canvas in a long triangle shape allowing the air to flow over the length of the cloth. It lifts the vessel as it passes. You must be warned, they require more effort to manage than your square rig. When the wind changes, you must adjust the set. If the difference is as minor as a degree or two, with this sail, there is more work to be done."

De Portu was not worried by the extra effort required. "We have a crossing of many leagues ahead of us. We'll need the advantage it will provide because there'll be headwinds all the way."

"Where do you plan to go, Captain? Are you chasing whales to the far side of the Earth?"

"At least to the far side of the ocean. Our world has been changing and the whales we hunt have been moving farther away each spring. They seek cooler seas, I think. If we are to follow the whales west, however, the winds will be blowing against us."

De Portu left out any mention of his real plans. At this point, fishing in Vinland was only for him and Shicki to know.

The talented artisan was also a shrewd businessman. "You'll need to switch the old one back in because those same gusts will most often be at your back for your return. Let me sew you a new square rig as a reserve."

The captain wasn't worried the old man might tell others of his plans, but still, he reminded him, "Please say nothing of this to any other captains. Not now and not after we are many leagues at sea."

"We are Basques, you and I." The old sailmaker spoke the words that didn't need to be said. "We are famous for keeping secrets. Nobody will know of this visit."

Chapter 19

Off Bergen, Norway

Approaching the crowded boat basin in Bergen, the first thing the crew could see was the port's seawall. It was built high to signal it was for serious working trading vessels only. Unfortunately, a whole armada of smaller day boats was already nestled alongside, seeking shelter from the gathering coastal winds. Fishermen from the little boats had strung their mooring lines tighter than spider webs in all directions. There was no room for new arrivals.

"Take us right into that tangle of little karves and tight against the seawall," Arne shouted. "We're bigger than they are." When he judged the boat was in the best spot, he commanded two boys on the wall to secure the longboat's lines. He never did notice the two karves the longboat swamped on the way in. Another was crushed against the robust masonry.

As soon as the boat was secure, Arne shouted to Dag, "You know where the fish buyers are. Tell me where to show them this basket of salted fish?"

"You don't have to go far, Arne." Dag pointed to a line of three-story wooden structures directly across the seawall. "See those colorful buildings all in a row? They're the guild's warehouses. The buyers dine on the top floor of the yellow one next to the town hall. That's where they decide who they want to deal with and what the price will be. All you need to do is go in and ask for a chance to sell our fish."

Arne picked up Aideen's basket. "I'm not *asking* for anything if what I'm holding is this valuable." He clambered up the seawall and crossed the cobblestone plaza to the row of wood-frame buildings.

The door of the building next to the town hall opened to a narrow staircase. The pungent odors of pine tar and wood shavings filled the shadowy gloom. He climbed to a second level, stepped into a darkened room with low ceilings, and shouted, "Who's in charge of buying fish?"

Two young men in greased overalls straddled a bench at the far end of the room, playing Hnefi. "You have to speak to the guild members," came the disinterested reply. One lad pointed to a second staircase. "They're having their lunch upstairs in their salon. They'll be up there for a while. You'll have to wait here." Arne turned and mounted the creaking stairs to the third floor.

At the top, a dim glow filtered into the salon through the sooty skylights, but the light seeping in had little effect on the dark timbered room. Upholstered furnishings confirmed these entrepreneurs enjoyed their comforts, and Arne could hear voices in the private boardroom and the clinking of silver utensils on pewter tableware. He set the basket of fish down, sank into a calfskin armchair, and planted one boot on the edge of a polished oak table. *They'll come out after they've eaten and see me here,* he decided. *I'll offer them this extra treat then.*

The food service ended, but the boardroom conversation dragged on. Arne continued to wait with uncharacteristic patience.

* * *

At length, four well-dressed gentlemen emerged. The tallest of the four had a full head of wavy white hair, a trimmed beard and delicate-looking hands. He immediately spotted Arne and stood motionless, not understanding how anybody would dare intrude on their space.

"We have an uninvited visitor who doesn't know his manners."

Arne stood. "I'm Arne Gunnarsson, and I know my manners when I need to know them … like right now." He pointed to the basket on the floor. "I suspect you've finished a fine smorgasbord of smoked salmon and such, but I brought you this treat, not as a compliment to today's meal, but more as an *entrée* to tomorrow's business."

To Arne's satisfaction, the fish dealers seemed to be taken aback by his controlled brashness. They didn't bother to identify themselves but did scrutinize the fish.

"And why do you think we would want your fish, Arne Gunnarsson? Perhaps you haven't noticed, but this warehouse and the others along this quay are bursting with fish."

Arne was ready with his response. "I see you already have a fine inventory of dried fish, but you have nothing to match this. I am offering you *cured* fish, a better product from Vinland. I promise it will change your operations forever."

The man met Arne's arrogance with a haughty attitude of his own. "You're part of that gang of rogues aboard the fat longboat we watched squeezing into the seawall. I can tell you straight away our guild is not interested in a distant fishery. The cost of sending things over there, compared to the value of what returns, makes the place a poor investment. In fact, Vinland makes no sense at all. That's why the colonists there gave up on it years ago. You should ask yourself this, Arne Gunnarsson. If we have all these fish here, why go so far to bring back stockfish?"

"Yes, Vinland is the other side of the ocean, and it might take a few weeks to sail there, but the catching takes less than one day. We could be there and back before your fishermen have hooked enough of these smaller cod you need to fill your warehouses.

And what's better, we'll do the salting on the return trip, adding more value."

Arne picked out a fish, cubed it into four bite-sized portions, and held them out, grinning. "Hungry customers will line up outside this old building after they taste the difference between the hard-dried fish you deal in and this salted product."

The man pecked at his sample. His reaction was immediate. "The taste is quite pleasant. Cured fish, you say, the same thing our cooks do with pork?" The others followed his lead, chewing and nodding in agreement.

"I already have all the salt we need for next spring. If you cover those costs and outfit us for next spring, we'll return with a load of cod, properly cured and ready for market."

The head of the guild made eye contact with his partners and responded with caution. "Please sit down again. You'll have to wait while we consider this." They ushered Arne back to his armchair, requested him to keep his boots off the furniture, and disappeared into their boardroom. Less than two minutes later, the foursome re-emerged.

"You have our interest, and you may have our money for the salt and to outfit one vessel next spring. Should the profit prove as solid on this cured fish as you say, there will be more money for more trips by our whole fleet. For this first one, we will place our company man onboard to collect navigational information and the precise location of this fishery you speak of."

Another spy to take along, Arne thought, but tried to appear appreciative, hoping the merchants would be convinced they earned added benefit with their stern bargaining.

"With the company's supplies up front, it's a fair agreement. We'll be back first thing in the spring." He turned to the staircase leading back down to the seawall, reminding himself he would not be letting the guild bosses in on the location of the Vinland fishery.

Back down on the seawall, Arne called out to ready the boat for home. "Stok, get the crew clearing the lines."

The crewmen struggled to untangle all the ropes. Several of the local fishermen became irate when they saw their boats being cast adrift. A shouting match erupted into a shoving contest until Arne took out his ax and chopped the last of the tangled mass, including all of his own ropes. "We won't need those at sea, and we can replace them at home. Now get your oars wet."

With a squall still blowing outside the harbor, Arne calculated the trip home to Stavanger would take the rest of the day. They wouldn't be able to tie up when they arrived, but that was a problem he'd deal with between now and then.

***

Once outside the protection of Bergen Harbor and beyond the many islands around its entrance, the sea conditions worsened and the crew struggled to keep the longboat's nose into the sou'wester, but Arne was enjoying the feel of the swells under his feet. He looked over at Stok and tapped his foot on the deck. "I knew right from the start this was going to work. Our boat has met the worst the sea has thrown at us, and I plan to tell every Norseman I see how we crossed all the way to Vinland and back on this fine, new boat and one that cost us nothing."

He pulled Stok off to the side and explained the details of the offer from the guild.

"If the guild insisted that we carry along a snoop, how do we deal with him when we want to sell the fish somewhere else?" Stok asked. "Will we do the same as we did with the priest we had to take with us before?"

Arne dismissed the issue. "Not quite the same, but I haven't decided yet how we'll make it work. Anyway, my mind is on another issue. We've been away for the whole of the season," he grumbled. "Our countrymen expect to see our new boat crammed with ivory, but we collected only these few coins in Limerick and a few more from Aideen. They're not enough to support the crew over the winter. And I can't forget,

I promised a double share to the men we left at sea off Ireland."

Stok piped up, "What you can't forget is there's always next year when we'll earn more than enough silver to share with the crew. Until then, we'll make up the shortfall with some of the stash from our hideout and replace it with what the guild pays us."

* * *

Pulling into the village wharf at Stavanger, the longboat's crew was given a wild welcome. As soon as Stok stepped ashore, he launched into ribald stories of how they crossed to Vinland and back and a shameful reason why they had nothing to tie the longboat up with. "And that's only part of the tales we have for you. There'll be lots more for the crowd tonight at our longhouse."

Arne shouted at Olaf, "You arse. Go ashore, get a coil of new hemp rope and secure the longboat to the wharf first and get Dag and Benji, plus two others to help me retrieve a few sacks of the reserve from the hideaway."

Olaf objected to letting Benji see where they hid their treasure, but he couldn't think of anything specific to support why Benji did not deserve Arne's trust.

Arne snapped, "Benji and Dag are both coming because the climb is steep. It's not as if you could haul your big gut up there, or Stok either with his peg leg."

The group took a small boat across to a lonely island in the middle of Stavanger Bay. Once on the beach, Arne headed to the top of a steep rise, shouting behind him to be careful of boobytraps as they bushwhacked through thickets to a bald outcrop. The sheer rock face was fractured in several places, and shadowy crevasses cleaved the bedrock.

"This is a great place to conceal your loot," Benji exclaimed. "You can see for leagues in any direction." He knelt and extended his arm into a dark crack in the rock. "Not there," warned Dag, "Arne has surrounded all those cracks with stinging nettles."

Arne goaded him on, "Go on, Greenlander, check for yourself." Benji stepped back.

The group followed Arne as he climbed up another knoll. After he surveyed the ground, they watched him slip into a concealed fissure. He tossed out four sacks of ornaments, tableware, and coins and told them to carry a sack each back down to the beach. He grabbed one pouch himself and followed them back down.

"I'm keeping this out for our two crewmen. If they do manage to walk home, they'll have earned the treasures in here and more."

Back in the brothers' longhouse in the village, Arne divvied up the coins and silverware. Each crewman agreed they preferred the long days and rough seas over the scars and broken bones they usually came home with. Arne enjoyed watching his crew bicker over one piece or another as he set the fire under an oversized crucible to melt down what they didn't want or couldn't use. Gambling began immediately.

Even Benji enjoyed the raucous scene. "My grandparents would be proud of how I've found something more interesting than brewing ale in the abandoned chapel, but I'm not sure they would approve of this."

Arne chuckled, "There's nothing wrong with stealing from the rich. It's too bad you're not making your dark ale though, Greenlander. It was tastier than the ale we make here."

* * *

The unexpected sight of two stragglers trudging into the village in the midst of a winter blizzard ignited an extended celebration. Arne whacked the bung out of a fresh keg of mead and announced, "Seeing you two is the final reward of last summer. Sit, drink and share the details of all the women you met and the pillage you stole on your long walk."

During the boisterous and drunken feast, the two unveiled their collection of loot, added stories of how they traded some for other goods and generally left a trail of mayhem from Spain to Norway. Their bawdy stories of the people between here and there kept Stok and the crew spellbound, but they admitted they hadn't expected it would be such a long trek. "We are thankful for the double share Arne, but don't ask us to do it again."

"And the priest?" Benji interrupted. "Did you drop him off in Aarhus?"

Both men took on a blank expression and confessed they left the priest 'somewhere in Spain.' Neither could describe where.

Benji jumped up. "Wait. You agreed to make sure he was alright."

"No, we didn't." Both men objected at once. "Anyway," one added, "he didn't want to come with us. You should have seen him. He kissed the dirt where he was standing, so we wandered off."

Later, Olaf complained to Arne about Benji. "I never did trust him, Arne. I told you before. And you heard how upset the little turd was when our men said they left the priest alone in Spain. Your Greenlander is trouble for us, Arne. Trouble."

Arne shrugged again. "You're not an owl Olaf, you're more of a seagull. If you're not flocking, you're squawking."

Chapter 20

Bergen, Norway, May of Year Two

All winter, cyclones relentlessly battered the outer coast of Norway. With nowhere to go and little to do in the chilly, wet weather, the brothers and their crew became as bored as all the farmers and fishermen in the village. Nobody was as rich as they were, but nobody wanted to get back to sea as much as they did. When the first spring wind blew in, Arne was quick to rustle up the crew and head up the coast to Bergen.

As soon as the longboat banged up alongside the port's robust gray seawall, Arne headed for the yellow warehouse across the square to remind the guild of their deal. A half an hour later a wiry, little man sporting a broad smile came out of the building and crossed over to the longboat.

The new man jumped down on the longboat with his sea bag and a cheery greeting. "Hello. I'm Sweyn. The owners reminded Arne that they had accepted his deal for a long-distance trip as long as he took a pilot along. I was quick to volunteer."

"Are you a real pilot?" Benji tested. "Arne says he doesn't need one. He uses his box compass, a sunstone, and then he sniffs at the air to be sure."

"That works for some, but the guild wants me to keep accurate records. It's the owners' money, so they insist on documentation. You have to understand, up to now there has never been proof of the amount of fish your captain claims you have seen in Vinland."

Benji was surprised. "Didn't Arne tell them how we can get 'two fish on a hook'? Didn't they believe him? I'm from Greenland and I've seen it. We had a priest aboard too. He named the bay for Notre Dame. You have to believe it was pretty amazing when a priest gave the place a church name, eh?"

"Here's the reason the guild wants me along." Sweyn dug into a leather sheath held securely on his belt. "I'll use this to get the exact location for them. It's the latest design of an old invention."

Benji leaned in with interest. "A mariner's astrolabe. You can work out your position with it by sighting the stars. My father studied to be a pilot before he was lost on a walrus hunt. Now I'll learn what he never was able to pass on to me. Will you teach me?"

"I'd be pleased to show you how the instrument works, and I'm sorry to hear your father wasn't able to show you, but he must have told you there's a lot more to navigation than this instrument."

Sweyn continued to sort out the rest of his kit, including a signaling horn, logbook, and a hand-drawn copy of a copy of the crude chart Leif Eiriksson made of the land around his Greenland colony. "According to this old sketch, it'll be straight across to Iceland. Next, we head southwest to the tip of Greenland. There will be long stretches of nothing but colorless sea and sky, so you can practice with this astrolabe then. And Benji, I've always wanted to see Vinland, so there'll be things you could teach me. Maybe you know why the settlement there was abandoned, for example. It would be good to visit Greenland too and learn useful tips for navigating among the ice bergs. That's all part of why I volunteered."

Benji shook his head. "I'll be glad to tell you, but it's already full summer there. Why do you think we call it Greenland? A few tall bergs may still be there in the shallows, but they'll be grounded on the bottom and easy to avoid."

* * *

The sky was clear and the winds light when Arne stepped out of the guild's building and strode to the longboat.

"Stok. These men are giving us all the gear, food and extra mead we need, so we're done here." Arne paced back and forth on the seawall, impatient for the supplies to be packed onboard. He shouted down to Dag. "See how I deal with the guild you said I couldn't deal with."

The smile lines at the corners of Stok's eyes grew deeper. "So we fleeced the church out of this fine vessel, and we departed Aideen's wharf loaded with her salt. Now it's free supplies from the Bergen guild. One final thing though. Tell me how you plan to cheat the guild after we are loaded with fish."

Arne was guarded. "We have things to talk over. Having this pilot along may be a disadvantage, or he may not."

Stok wanted more details, but seeing the final load of supplies packed aboard, he dropped the steerboard into the longboat's yoke and swung the tiller hard over. The crew took long sweeps with their oars and the longboat came away from the wall. Arne took the tiller from Stok and grinned as the longboat shunted and bumped smaller boats out of its way. Clear of the inner harbor and heading out Bergen's long channel, a breezy nor'easter greeted them. The longboat picked up speed and Arne's grin broadened.

"Ship your oars men. Lash them under the gunwales. If we hold onto this nor'easter and ride the currents, we'll fly right to Iceland. No stop at the Faroes on the way either."

Sweyn returned to his copy of the old chart and scribbled numbers in the margins. "Twelve days to Iceland, if I calculate right. And on to Greenland. I'm looking forward to a stopover there."

"No need for a stopover there," Arne said. "Thanks to your guild's instructions, we have all the supplies we need." Arne gave his usual impatient wave. "That's the plan, Pilot."

"No visit to Greenland, too bad. I'll plot a direct route then. It will be Iceland and on to Vinland."

* * *

Under a welcome stretch of clear skies, the longboat took full advantage of the steady nor'easter. Arne found a swift current to carry them along. On the tenth day, the low volcanic peaks of Iceland came into view.

"I am pretty familiar with the approach to this harbor," Sweyn said. "Do you want me to take the steerboard for the run in?" Stok stepped to one edge and Sweyn effortlessly guided the boat around the island's southern peninsula, headed straight up the shore and slipped neatly into Iceland's busy boat basin.

"Hey Stok. Do you see all the puffin colonies over there?" Sweyn pointed to countless groups of small birds that sounded like piglets as they flitted around the rocks in their noisy mating dances. "The Icelanders say there are more puffins flying around this island than there are people in the whole rest of the world. I ate two for breakfast with my porridge last time I was here. It was the puffins I ate, not Icelanders."

Arne studied Sweyn. "On our last trip to this harbor, I wanted to eat the Icelanders when they wouldn't trade us any walrus tusks or fish. What brought you here?"

Sweyn blinked nervously. "I'm surprised you didn't figure that out already. The guild managers have a closed network of suppliers, including this one. We hold a monopoly on all the walrus tusks and fish traded in the Norse countries. It's the same for the Faroe Islands too. I was sent here to set up the original deal by those same men who outfitted you for this trip. They treat Iceland as their private island. Did you think nobody told them when you visited last year? And if you are able to open up a profitable fishery in Vinland, the guild will be quick to grab the market rights to the fish caught there as well."

Sweyn could see his confession was completely unexpected. Arne's nostrils flared as he attempted to mask his surprise by gazing at a colorful little bird struggling to lift itself up off the surface.

Once the longboat was secured, Sweyn offered to lead the crew to the one inn. "You'll all drink for free here men. The guild owns the inn too."

The crew cheered. Arne declined to join them.

**

On the short walk over to the tiny inn, Benji caught up to Sweyn and asked if he had some time to talk about navigation. With two jars of ale and a table in the corner, Benji asked, "I know how to use an astrolabe to figure my position, but how do you set a bearing for somewhere if you're not sure how far away it is ... Vinland for example? We don't know how many days it is from here."

"You're right, Benji. When moving into uncharted places, it's not enough to follow a compass. With an astrolabe, we follow the proper latitude, that is a line across the sea. We calculate the number of degrees up to the north or down to the south from the line. The other thing we need is how far along the line we have traveled. The men who control the guild want to know, so they sent me to come back with reliable answers."

The next morning the crew was hungover, but Arne stomped up and down the deck, disturbing their sleep. "Ten more days to the fish," he announced. "Time to go."

Sweyn stood up. "I figure we'd need a dozen days Arne, but this vessel moves along faster than most others I've seen. You can turn the slightest gust of wind into an extra league or two of speed, so I think ten days may be right." Arne ignored the compliment.

On the eighth day out of Iceland, steady winds still pushed the longboat along, giving the rowers little to do but play dice or sleep. Benji watched as Sweyn slipped his astrolabe out of its pouch, adjusted the dial and made notes. The experienced pilot worked through each step again, frowning.

"What's the problem, Pilot? Is your new astrolabe not working?"

"Arne must have found another surface current. According to my readings, we've already come south of fifty-three degrees latitude. We might be sighting Vinland by tomorrow."

Benji scanned the horizon around them. Numerous seals poked their whiskered snouts up through the chop. "There," he pointed and shouted to Sweyn. "Maybe we're near Vinland or maybe not, but those fish-eating monsters are a good sign we're near land. They're out here because the eating is good."

Arne heard Benji and shouted to the crew, "Ready your handlines and hooks. Remember when Benji told us there'd be fish underneath us? He was right. But don't forget to let him bring up the first one. That way he'll have to clean the rest of it for us." Benji had the cork off his hook and his line descending to the depths before the rest had rooted through their kitbags.

"Hey. There's a plank sticking out down there," he shouted. The others all moved to the rail and leaned over. Nobody saw anything and accused Benji of making a bad joke.

"No, no. I'm serious. There's a loose plank. With you all over on this side now, it's gone down below the waterline."

"Get back over to the far rail, all of you," shouted Arne. "I'll stay here and check the hull when this side rises back up. Benji, you go below and check if it's the one you busted out last summer off Ireland."

Arne leaned over, spotted the protruding board and waited for Benji to re-emerge.

"Yup. It's the patch. It's loose and sticking out," Benji explained. "It's not leaking too badly because the plank is still in place."

Sweyn butted in, "Arne, if we're at Vinland already, could we stop at the old Norse settlement on the northern tip? It was marked on my chart and I should check if the location is correct. Once we are there we might find more of the extra-long rivets and disks the boatbuilders use to refit the plank and the patch, too."

"Set a course to put us close," Arne ordered.

* * *

The clouds hung low in the evening sky and the dim light scarcely illuminated a rust-toned shoreline. Only stunted tamarack trees distinguished it as land from the gray sea around it. Sweyn's first impression of the storied site was a let-down.

"There was supposed to be a whole village here once. If there ever was, it wasn't here long. And I don't see any real trees to trim up for plank repairs either. You think this is the actual settlement mentioned in the sagas Benji?"

"There must have been tall trees here once," Benji said. "They must have hacked them all down for lumber or firewood maybe. Our fishermen claim there is a shallow inlet leading into the abandoned settlement. We should see their longhouses as soon as we're closer. Maybe there are still a few spare timbers left in the lodges. We could scrounge them for repairs."

Stok lifted his steerboard up out of the yoke and pointed. "Arne, you want us to keep pulling hard and stay in the center of the inlet?" Arne nodded and shouted at the crew. "We'll run her up onto the shore. But remember, this boat is not flat-bottomed like our old one. Line up on the other side of where the plank is loose. When we bottom out, jump off. Once your feet hit the beach, support the hull as the tide falls and tip us to one side. Have the side with the loose plank facing up. After, search the site for drinking water."

The men pushed on one side of the hull to expose the patch over the cracked plank. Sweyn slogged through the seaweed to survey the rest of the longboat's hull. "There are a few other loose timbers too Arne. They're not damaged, but they are loose. I think we can pin them back tight against the ribs. We'll need a little oakum to force in between the planks and pine tar to go paste over top."

"Sweyn, just lash the loose planks tight to the ribs. I want to keep moving as soon as possible." Arne was proud of his ability to command crewmen and navigate at sea, but he never bothered to learn how to build, maintain, or fix any problems with his vessels. Sweyn shook his head and changed the subject.

"You should have the crew shove her over to check the other side, to give us an idea of any other issues we have to deal with."

Arne ignored him. "As soon as the one leak is fixed, we'll need to locate some fresh water and maybe a few seabird nests with fresh eggs. Then we leave on the next tide, right?"

Sweyn didn't respond. He joined the crew as they broke into teams of two or three each to crisscross the site.

* * *

Benji and Dag traipsed across the eerie landscape and within a few hundred paces, stumbled right into the collapsing structures that the settlers had abandoned. Rows of sod foundations gave way to separate plots of tilled soil. The business end of a broken hoe protruded from the soil where vegetables had once grown. Grasses had grown high and obscured the edible plants. A frightened ptarmigan squeaked a warning to its unseen brood and fluttered away into the ground fog.

They followed a well-worn path leading to the longhouse where a wooden bench, supported by only three legs, stood sentry by the main entrance. The sod roof had long ago fallen onto the stoop.

"This is a bleak place, Dag. There's nothing but sea on three sides and a dreary mist covering the earth like a felted blanket."

"Can you imagine living here, Benji?" Dag asked. "This breeze is chilly and there's barely any firewood. Even right here where the longhouse is, the ground is swampy. That must be why the single men gave up. They had no unmarried girls to snuggle up with during the dark nights."

Benji chuckled. "It's the same where I live in Greenland. We do have trees still standing around, and it's a pleasant place to raise a family, but there aren't any unmarried girls to do that with. That's why Grandpapa wanted me to join your crew ... so I'd find a girl over in the old country and take her back to Greenland."

"He had the right idea," Dag agreed. "I think I'd do the same."

* * *

Benji and Dag hiked back to the beach and reported to Arne. "No usable wood anywhere. They must have had a well, but we couldn't find it. We did see a tangled grove of wild grape vines though. Maybe they only drank wine."

"There was one winding little stream," Dag interrupted. "But the flow was slow and all cloudy with a skunky stench. It didn't look fit for a hound. I drank some to test it for the rest of you and I'm still standing though."

Dag staggered, gripping his throat and fell over, pretending to be choking to death. He waited for Arne to react, hoping the performance would lift his captain's mood. Benji laughed, but Arne ignored them both and called the rest of the men back. "The boat's careened over so far, the deck's not level enough to sleep on. It's going to rain all night so we need to find some shelter."

"This place is depressing," Sweyn said. "I'm sure the ghosts of the settlers are still floating around. I'm not sure I want to stay the night, but I did see part of the roof is still on their old foundry. We could shelter there and maybe dig up enough iron fittings to make the plank and patch tight."

"Collect your sleeping mats from the longboat then," Arne ordered. "We're going to spend the night in the old smithy. While we're there, check for any short timbers we can use for repair. We'll leave buckets out overnight to collect some rain. If there's enough, we'll drink some and boil the rest for what fish we have left onboard."

***

Arne's sleep was fitful. Before the first light warmed the eastern sky he roused Sweyn. "You're supposed to know what you're doing. Check the planks and the patch. If they are tight, get the crew to push us back off the beach. It's time to leave." The rest of the crew scurried down to the boat and shoved the hull as best they could. When the boat was barely afloat they clambered up the side and onto their benches. Arne dropped the steerboard back into its fittings and swung the tiller. "Back four rowers, pull us out of here. We have one last day before we fill our holds with fish."

"Is it all that simple for you, Arne?" Sweyn asked. "Navigating, fishing and quick repairs, I mean. You should consider this place. It was built with the hard effort of your fellow Norsemen, but it didn't work. We can see it was a mistake and why they abandoned it. Doesn't this settlement remind you of the choices we have in life? I mean, can it all be so simple?"

Arne stared at his pilot for a moment then nodded. "You're right. Life is not simple. We should get rid of more things that don't work."

Chapter 21

Notre Dame Bay, Vinland

Arne followed the near shore for most of the day, expecting to see the small islands and the bay where they met the Greenland fleet last year. Tired of squinting at the shoreline, he announced, "This has to be the place. Those islands out there may be different than we saw last year, but this is the same bay as before. 'One Day Bay', that's my name for this place because a day is all we'll need to load up."

He rushed back to the steerboard where Stok stood. "Let's head over there nearer the shore. We need to start fishing."

Sweyn listened to the commotion then held his astrolabe up so light from the sun shone down onto the sighting loop. After several calculations he asked, "Arne, are you sure this is this the bay you promised the guild we'd find? Because my reading gives me an exact fifty degrees north."

Arne walked over to Sweyn and couldn't resist boasting. "It doesn't matter what number of degrees you get, just count the number of fish. The men you work for will soon have proof of what I told them."

Benji chimed in. "Fifty degrees? Is that where we are? Our priest was excited by how fast we caught the fish last time and kept saying, 'The church must be told the exact location. The church must know.' Tursk told him sixty-three degrees, so he must have lied, because the Greenlanders don't want more boats coming over to Notre Dame Bay."

Arne's reaction was quick, "It's not Notre Dame Bay. We're calling it 'One Day Bay' and nobody will reveal its location."

"For you, maybe," Sweyn shrugged. "I need a reliable position for where the fish are for my report to the guild. Leif Eiriksson didn't make good charts with modern instruments."

"You don't need an exact position," Benji added. "There's more codfish swimming in these parts than stars twinkle in the night. Remember what you said about the puffins in Iceland, nobody bothers to eat a puffin here either, because there's a big, tasty cod waiting in the water." Sweyn ignored the jibe about having eaten the little birds.

Arne put his hand down to his belt, gripped the handle of his ax and gave both of them an icy stare. "And you'll both end up down there with the fish if you don't keep all this to yourself. The rest of you, start filling this boat."

* * *

As soon as the crewmen slipped the corks off their hooks, the boat deck became a slippery mess of flipping fish. The crew couldn't keep up with the numbers coming aboard until Arne shouted, "Only do the fattest ones. Throw the little ones back, the same as the Greenlanders do." Clouds of noisy gulls filled the sky above them as soon as the words left his mouth.

The first time two fish came up on one hook, Sweyn turned to Arne. "I've never imagined this before. The guild should have believed you."

As the sun climbed higher in the sky, it shone straight down on the open deck. None of the men had cover from the heat, and most stripped down to their woolen breeches.

"Benji," Arne called. "We're all sweating here. Where's the drinking water? I told you to fill the kegs."

"We planned to refill the kegs at the abandoned settlement, but the water there tasted worse than mouse turds, remember?"

Arne found the expression amusing but he knew the crew needed fresh water. "Somebody has to go ashore. It might as well be now."

Stok called Arne back to the stern. "We do need to fill the water kegs, but we need to deal with Benji too. Olaf told me the Greenlander was seen scheming with our Irish friend Aideen before we left her place last year. Olaf thinks Benji told her we planned to keep her salt and sell this fish in Limerick. And you remember how upset

Benji was to hear our men left the priest alone in Spain. If the church hears what we did, it could be a real problem for us. And you know what's worse? You let him see where our cache is buried on the island. Bad, Arne, bad."

Arne spat. "Yeah. So do something then."

Stok leaned over the rail. "If you send Benji ashore, he could go with Olaf the *Ugle* and one other. Three men go ashore — two come back. Olaf will make up a frightening story of how the *skræling* captured our Greenlander. Problem solved."

Arne nodded and turned away, but Stok grabbed his elbow. "And now, there's Sweyn? Since we don't plan to let him in on this deal, we better plan how to keep him out."

Arne smirked. "You let Olaf deal with Benji, I have a place in mind for Sweyn."

* * *

With their holds full, Stok told the rowers to ram the pebble beach head-on. He called Olaf over and pointed to two empty barrels on the deck. "Pick a crewman to go fill these full, anyone but Dag," he insisted. "And there's one other thing. I want you to invite Benji to go with you, then leave him there in the forest."

Olaf stared straight at Stok. "Huh?"

"Benji will stay ashore and live with the *skræling*. He doesn't know it yet, but he'll figure it out after you've left him deep in the woods. Maybe the *skræling* will find him, maybe they won't."

Olaf hesitated until Stok handed him a pair of leather thongs. "Before you come back, tie Benji to a tree far in from the beach. These will keep him where you put him until we're leagues away."

Olaf smiled. "I can arrange to fix the maggot. He thinks he's an expert on fish and seabirds. Let's see if he's an expert on untying his hands from behind a tree without a sharp blade."

Stok called over to Benji, "Go ashore and help Olaf the *Ugle* fill the water barrels. Maybe you'll see the little friends, the ones the priest met before." Benji ignored Stok's comment, jumped over the side and stood knee deep in the breakers waiting for the empty barrels to be lowered down. With Olaf and one other crewman, he set off to locate a fresh stream.

* * *

Olaf trudged back down the shore, rolling a full barrel in front of him. The crewmen onboard the longboat spotted him and heard him yell, "They took Benji. The *skræling* just took him." The second man was steps behind Olaf. He dropped to his knees panting and pretending to be running scared.

"There was nothing we could do," the second man added. "The *skræling* had long knives at us and there was one with an ax — a broad, sharp ax. Olaf and I grabbed this barrel and took off faster than a pair of scalded rats."

Stok shouted, "Hoist the keg aboard. The rest of you get your oars ready. We better get out of this bay before we're boarded."

"No. Wait. You have to go back," Dag demanded. "They won't put up a fight. And what if they do? You've got our whole crew here. I've never seen you back down, Stok. We can rescue Benji."

"We can't chance it," Stok insisted. "If they have sharp knives, we could all end up as fish food."

Dag drew his dagger. "This is sharp. We all have one." He waved the blade right at Stok's chest. Stok jumped back, drew his own knife and pointed it at Dag's face.

Sweyn stepped between them and shouted, "Back off, Stok. Dag's not the problem here. Anyway, he's right. Olaf's story doesn't make any sense. Where would they get real weapons? All they'll have are pointed sticks. We should go back and make a deal with them. You can't abandon the lad."

Stok backed up another pace. A few of the crewmen drew their knives and prepared to go ashore until Olaf insisted, "I'm not going back in there."

"Not me either," agreed another crewman who was in on the hoax. Arne let the turmoil continue long enough for the standoff to firm up. Trying to sound wise, he looked at Sweyn and said, "No, we can't risk a battle and losing the men we need to get your fish back across to Bergen. If the *skræling* want Benji to survive, they'll help him. He'll probably make some of his brown ale for them. If the *skræling* don't find him, the fishermen from Greenland will come along some day. He was always saying he wanted to go back to Greenland, right?"

Dag looked at each of the others. Nobody was moving. "At least we can stay here in the bay. We're safe on the boat for as long as it takes to pack the load in salt. Benji will escape soon enough. We'll see him running out from those trees. He'll swim out to us."

Arne spoke up again. "We're leaving. We'll salt the fish as we go."

Dag stood at the rail, knife in hand, mulling over his options. *I could jump over and find him, but by myself, what could I do?* He could see the rest of the crew was already settling on their benches, preparing to row.

Stok sheathed his knife and joined his brother at the tiller. Folding his arms, he whispered, "We won't be seeing our Greenlander again. There's another one of your itches you won't have to scratch. Now for our navigator from Bergen. He might have to be left here too."

"I told you," Arne snapped. "I have a different place in mind for him, on a different island."

***

For the next three days, Arne did little but stew about Sweyn. He had deserted two of his own men, the priest, now Benji. *To keep a clean boat, some things have to be flushed overboard*, he kept repeating to himself, but it was becoming hard to live with the idea of getting rid of men like fish guts.

Also gnawing at Arne's ego was the threatening gesture Baldr made last year at his fishing bank in the Faroes. Arne kept telling himself, *There's got to be a way to put one problem against the other*. On the fourth day at sea, an idea came to him. He called Stok aside.

"Baldr and Sweyn are like rivets, the same as the ones we used on the loose plank. Alone, they are just sharp points, but if I connect the two they become a fork. Remember when we were in Ulster and that monk stuck a big silver meat fork into your leg? He made good use out of it. Let's head for the Faroes and stick our fork into Sweyn and Baldr. Except it won't be a meat fork. It'll be a fish fork." Stok nodded at Arne, but he had no idea what his brother meant.

Arne shouted to Sweyn. "Pilot. Work out our position and plot a new heading. Stok wants to stop at the Faroe Islands. It's not far off the compass bearing for Bergen. He wants to see his lady and the crew need another night of drinking in the public house in the village."

"It's not much of a reason to change direction Arne. You know there's no walrus ivory for you, and the guild cornered the fish trade there long ago. You can't sell this fish until we get home so what's in the Faroes for us?"

Arne nodded and turned to stare out to sea. "People think the schools have moved away, right? Well, they are wrong. There are cool currents still flowing south of the Faroe Islands, neither you nor the guild has been told about, and shallow areas too, with cod waiting to be fished."

Sweyn waved off the comment and continued to work on the course change.

Later, Arne asked Sweyn, "Did you hear what I said? There's a place in the Faroe Islands with twice as many fish as anywhere else. An old Viking named Baldr told me."

Sweyn was blunt. "Arne, I've been to the Faroes enough times to know there's no such place. I'd be the first to know if there was. Bergen has a monopoly on landings from all the fishing grounds and the guild there trusts Baldr. I know him too. He's a reliable man."

"Is that right?" Arne sniffed and blew his nose over the rail.

Chapter 22

Notre Dame Bay, Vinland

More than ever before, Benji noticed how the leafy tree canopy absorbed most of the daylight. Rays of sunlight barely filtered through. The forest was humid and airless. It seemed to swallow him. Nothing interrupted the silence except the sound of his own breathing.

He wasn't surprised by what happened. It was clear Olaf didn't trust him, nor Arne, but he hadn't expected Stok to lead the plot to abandon him. Questions kept running through his head. *Did they leave without me? And Dag, where was he when they left? What do I need to do first?* He checked his belt for his blade. *They took it. No surprise.*

With his hands tethered behind him around a lofty black spruce, he struggled to loosen the binding. Its scaly bark chewed at the skin on his wrists. The more he struggled, the more skin was scraped away.

It took him much longer than he expected to wriggle free. After struggling to loosen the thongs, the shredded skin on his forearms and wrists stung. He leaned back on the tree with his legs bent and propped his elbows on his knees, burying his face in his palms. The tree bark scratched his bare back. Sweat dripped from his forehead and mixed with the blood trickling down his forearms. Red splotches dotted the waistband of his dirty-white wool britches. There was no doubt in his mind that he would survive, although it was going to be harder without his blade.

The late afternoon air was humid, much more than anything he was used to in Greenland. He stood and looked in several directions, trying to get his bearings. The heat made him sweat and he was attracting flies. They swarmed him and attacked his exposed skin. To get away from them he bushwhacked his way up to the top of a hill, hoping for a stiff breeze to blow the flies away and, more importantly, to see if he could spot the longboat.

Dead spruce branches obstructed his path. They tugged at his britches and scratched his arms. Low bushes tangled his ankles while thick clouds of mosquitoes continued to torture him everywhere else. At the brink of a high hill, light gusts did blow the insects away, but sweat still blurred his vision as he tried to get a wide view out onto the bay.

A small bark canoe with two figures floated in the distance. "Am I seeing things? Dag is coming back for me." Tears welled in his eyes and slid down his cheeks. The paddlers had their heads down, gathering the fish dumped overboard from the longboat. "Dag, Dag." Waving frantically, he shouted and jumped. There was no reaction from the canoe. "No, it must be the Forest People."

Stepping over the brow of the hill, he half-ran, half-stumbled down the loose rock slope right down to the shore. Clouds of insects appeared again and pursued him aggressively. As soon as he was knee deep in the surf, he dove below the surface to escape the insects. He held his breath as long as possible, poked his head up, and swiveled around. "It's still there. HOY, over here." The two figures in the canoe turned, studied Benji for a long moment, and looked back at each other.

"HOY, yes. Come this way," Benji shouted, saltwater swishing in his mouth. When the canoe turned and headed for him, Benji had no idea what to expect, but made for the shore thinking, *What will they think I was doing here? Maybe they'll be angry that I was on their land. Will they make me their prisoner?*

* * *

Surfing on an incoming wave, the canoe plowed through the floating seaweed and crunched into the stony beach. The men stepped out and stared at the stranger before them. Benji couldn't think of anything to say. He wanted to thank them for coming to collect him and extended a hand. "I'm Benji," he said, trying to keep his voice from cracking. They both noticed the oozing scars on his wrist and retreated.

"No, no. Don't worry," he pleaded, tapping his bare chest. "I'm Benji Leifsson. I was on the longboat." He pointed out into the bay. "Did you see us out there? There was a whole boatload of us. You must have seen us. I know you hide from fishermen, but you must have been watching as we went by." Nervous energy was overtaking him and he rattled on until one paddler put his hand up as if to say, 'Stop.'

They signaled for him to step into the center of their canoe and sit among the fish. The breaking waves were high, but with a quick maneuver they pushed the boat off leaving a rut in the gravel. They headed their canoe for the mainland until spray began to break over the rounded bow. Recognizing the problem they slowed and shifted their positions to make the canoe's bow lighter.

Benji remained as still as he could, squatting in a shallow puddle in the bottom amid a mass of loose fish. He stayed still, but he couldn't stay silent. "You know, I was here before. Our priest was talking to a pair of boys and …"

The paddler in the bow twisted at the waist and spoke to the one behind, continuing the conversation in brief snatches. "I know they are talking about me, but they never look at me." It made Benji more worried about his fate.

"Am I going to be your prisoner now?"

There was no reply, but there was no sign of malice or tension in their voices either. He took that as a favorable sign and settled. "You have no idea what I'm saying, do you? And I can't understand you. I guess I'll have to wait and see what happens next."

As the canoe approached the shore, Benji could make out a cluster of lodges. Smoke curled above a handful of cooking fires and spiraled into the clear sky. The whole village must have been alerted as people continued to walk out onto the beach. Parents corralled the numerous children. Benji pictured how his grandmother would do the same if the situation were reversed. *She'd warn me never to trust outsiders.* A feeling of emptiness filled him. *I'll not see Grandmama again. I'll never have a chance to make things right for Aideen.*

Paddling with care to keep their overloaded canoe stable, the two rescuers and their apprehensive passenger covered the rest of the distance to the shore without mishap. The boat grounded out in front of the gathering crowd. Nobody said a word.

Benji stepped onto the sand. He didn't feel threatened, though the crowd was staring intently at him. The two paddlers hauled their canoe farther up out of the surf, collected their fish and walked off. Nobody else moved.

Chapter 23

Tórshavn, Faroe Islands

Arne and Sweyn avoided each other for the rest of the trip to the Faroe Islands. At the pier in Tórshavn the longboat was secured and Arne told the crew to head for the public house. He stayed behind, huddled in conversation with Stok.

Sweyn joined the crew on the walk to the ale house and quickened his pace to get closer to Dag.

"Dag, you have to tell me, what's going on? Arne insists there's some unknown fishing grounds somewhere, and he's acting as if I'm not going to like it when I find out. What are he and Stok back there scheming about? Can you tell me or are you going to fall in with them after what Stok did to Benji? Benji was your friend. You should see things clearly now because you could end up the same as he did."

Dag was slow to answer. "Some of the things going on I can't tell you about."

"You know you're not the most favorite crew member, Dag. We should stick together and find out what those two are cooking up."

Dag stopped in his tracks and turned to face Sweyn. "Let's talk at the tavern. We can sit in a quiet corner and pretend we're playing Hnefi or something."

At the public house, Sweyn asked for two mugs and the Hnefi board.

"You can't use it," explained the bar wench. "The most important game piece is missing. Some old Viking was playing with it a while back. Now we can't find it."

"Give us all the pieces you have. It'll be all we need." He pointed to a vacant table in a darkened nook. "We'll set it up over there."

The barmaid bundled up what pieces she could find in a basket and passed them over. "I guess it doesn't matter if there's a figure missing because it's too dark over here to see what we're doing."

"Right." Sweyn took the game board, spread it out on a stool, and placed the pieces in a pattern as if the game were underway.

* * *

The two pretended to be playing the game until Dag whispered to Sweyn, "They're not going back to Bergen."

Sweyn's voice rose. "But they have to."

Dag touched his finger to his lips. Sweyn lowered his voice and continued. "Arne's got an agreement with the guild. And after all the fish I saw, I'm sure their deal will be renewed for many seasons to come. What's the problem?"

"Well, what you don't know is how Arne expects to get a better deal in Ireland. And anyway, he's tired of going back and forth to Vinland."

"Wait. You mean he's not going to Bergen," Sweyn exclaimed. "Nor to Vinland. The fishing is amazing there. Where will he get the fish for this new market?"

"The locals here in the Faroes have a fishing ground they keep to themselves. It's a special place the guild has never been told of. An old Viking showed Arne last year."

"What do you mean, a special place? There's nowhere around here that the guild hasn't heard of."

"This old Viking, his name is Baldr, he knows the whereabouts, and he says they have kept it from the guild. Arne and Stok plan to fish there whenever they want to. They don't need to go back to Vinland. And they don't need the guild either. They plan to take it to Ireland."

"So, it's true what Arne was saying. Baldr is the guild's lead man here in the Faroes. If he's making deals on the sly, the guild will make it hard for him. And you won't be welcome back in Bergen if they connect you to some side deal Arne is making."

"I don't care now. I'm telling you this because I'm not going back to Bergen. I plan to go to Greenland. Benji is probably there already. I'm sure he survived in Vinland and got back home with the Greenland fleet."

"You're going to Greenland? How will you get there?"

"I plan to hide out here until Arne has gone. There's a mission church on the other side of this island, St Olav's. I'll hike over there tonight and ask the priest to take me in. Hopefully, there'll be a coastal trader to take me to Iceland. In the spring, I'll locate another trader going over to Greenland."

"What will I say when Arne connects the two of us?"

"That's why I am sharing this with you Sweyn. Arne will have to think twice about how to cover his tracks when he finds out more people know."

Sweyn sat back on his stool, trying to figure out his next move, when he heard heavy footsteps right behind him.

* * *

"Sweyn. What are you doing here today?"

The anger in Baldr's voice startled Sweyn. Before he could think of how to respond, Baldr blurted out, "You're the guild's man. Are you with Arne? His longboat and those thieving raiders are back at the wharf."

"Yes, Baldr, yes, I am. The guild sent me along with Arne to confirm the location of the fishing grounds he had found in Vinland. We're on our way back with a full load. The guild will be satisfied that their funds were well spent, not to mention fascinated with my information on the new fishing bank."

"Why don't I believe you? I'll tell you why. Arne is a schemer and the guild wouldn't do business with him. If you're with him you must be in on his schemes."

Baldr turned and stomped out, but not before he kicked the stool and the Hnefi pieces scattered noisily across the floor.

Sweyn gave Dag a worried look. "I'm not sure, but I think we should warn Arne that Baldr is not happy with us being here."

"Just another reason for me to disappear," Dag responded. The two left the tavern, but before they returned to the longboat Sweyn pulled Dag into a dark alcove between two buildings.

"When we get there, I'll keep Arne busy. You sneak around, get your gear and disappear."

At the wharf, Sweyn spotted Arne still deep in conversation with Stok. He shouted, "Arne, I know what you two are planning. The guild is not going to be happy."

Arne chuckled. "It took you long enough."

* * *

Before Sweyn could say anything more, Baldr marched up to them with a gang of local fishermen, and Baldr spat the words out. "I told you the guild controls all the trading on these islands except those fish. And they are for us Faroese alone."

"But Baldr, you're a Viking," Arne shouted back, "and as a fellow Viking, I came to warn you."

"Warn me? Warn me of what?"

"We have the guild's pilot with us. He's been with us in Vinland.

"We know Sweyn. He's been here before, but he doesn't know anything we don't want him to know."

"We've already let him in on your private fishing bank, so I figured you might want to deal with that problem before he tells the guild."

Baldr climbed up on the deck, now eye-to-eye with Arne. While all eyes were on the two men, Dag and Sweyn slipped away unseen and headed for St. Olav's mission on the other side of the island.

"Why would you tell him?" The blood vessels in Baldr's neck bulged, making his tattoos ripple. "What do you get out of this?"

Arne stood, stony-faced. "For me, I think of this situation as a fork with two tines. Let's call it a fish fork. Now, say Sweyn was to go back to Bergen with us and tell his bosses where your private fishery is. The guild finds out, and your secret is out. Since I have a firm deal with the guild, I would be allowed to fish there. That's one tine on this fish fork of mine."

Baldr sputtered and gripped the handle of his knife. Arne ignored him and kept talking.

"But suppose you take care of Sweyn for us. Keep him right here so he doesn't ever return to Bergen and inform the guild. The thing is, either way, my crew is going to show up on your private grounds whenever we want to. That's the other tine on the fish fork."

Baldr couldn't think of a response. Arne gloated and added one last detail.

"Oh yeah, and there's our crewman, Dag. I don't know where he went, but he's from Bergen. If he ever gets back there, he might tell the guild. You'd better keep him here, too. We're going to fish out on your grounds whenever we want, but you do see how bad this could be for you if Sweyn and Dag ever get back to Bergen."

Baldr spat on the deck.

Chapter 24

Dingle, Ireland, Easter Sunday

Aideen spotted the silhouette of a fully rigged tall ship in the distance. It was cloaked in fog, floating like an apparition above the horizon. She had been fooled by mirages before, hoping for the return of the Basque captain as he had promised. *More mischief,* she muttered, *like my dreams of fish swirling around the pillars of my wharf.* But when she checked later, the fog bank had lifted and she saw this ship wasn't moving on. It was closing on her shore. She ran to the edge of the dock, stripped off her wimple, and waved it into the air as high as she could. "It's Captain de Portu's caravel, *Espiritu,*" she shouted, "and it's heading in."

Tacking and jibbing, the tall ship in the light breezes seemed to take forever, but when it did nudge up against the wharf, she recognized de Portu's familiar tan waistcoat and faded beret. She sprinted up the gangway to greet him and, without thinking, planted a kiss on both cheeks.

"What an affectionate greeting. Thank you," de Portu exclaimed. He raised a hand to mask his thinning hairline and said to himself, *She does fancy me. I am not too old for her after all.*

"*Kaixo*," she said. "I hope I remembered correctly how Basques say Hello."

"Yes, yes. You speak Basque like a native. And speaking of our Basquelands, I have remembered what I promised during my last visit."

Aideen eagerly scanned the deck for any of the rotund barrels of wine she had long been expecting. There were none.

De Portu slid a hand into his waistcoat and pulled out his flask. "Here is a gift of our fine rosé, as promised." Aideen cocked an eyebrow. The flask went back into its pocket. As if on cue, Shicki appeared with a single cask propped up on his shoulder.

At that instant, a tall figure in black clerical garb stepped up onto the deck. Aideen jumped back.

De Portu chuckled. "Ah ha. And who is this with me, you wonder? A fully ordained priest? I remembered how you must avoid the import taxes imposed by your vicar. The fully ordained man here has accompanied us to ensure this keg has been suitably blessed. This juice of the grape could be your gift to the local parish for the celebration of communion. Appropriate, no?"

"Are you serious?" she demanded. "As if I don't have enough to deal with. This priest could be the end of me."

"I ah, I ah …" De Portu shook his head and began again. "I regret my mischievous trick. Perhaps it succeeded better than intended. Let me explain. He asked if we could carry him from the Basquelands closer to his home parish in Denmark. He was a passenger, is all."

Niels stepped into the conversation. "The captain is correct, Sister. I am but a traveler and one many leagues from home. My sole interest is to continue straight there. I should make contact with the head of your local parish first, however. Could you tell me where your bishop resides? I'm trusting he'll know of a caravan heading to England for me to join. I'm sure there'll be a ship from there bound for Denmark."

Aideen jumped at the chance to speed the priest on his way before he learned her plans for wine deliveries. "As a priest, I suspect you might wish to deliver the *communion* wine to our Vicar Maurice yourself. To assist you with this cask of communion wine our captain has brought us, let me offer you my mule and cart. To be honest, it's his mule and cart, so it'll be good for him to see it serving parish interests."

The wine cask was secured on the cart, and Aideen hitched up her mule. "This being Easter, Vicar Maurice should be busy feting his servants, as is the tradition on this one day in the year." She gave Niels directions to the tower house and thwacked the mule's rump. "Let him lead the way."

"Thank you, Sister. Your assistance is most appreciated. Oh, and don't fret. When

I do meet the vicar I won't mention the balance of the shipment of *communion* wine the captain stashed below decks for you."

* * *

As soon as the cart carrying the young priest was out of sight, Aideen turned back to the captain. "You have let me down, Captain, and exposed my entire business to a priest. 'Tis sure he'll tell the vicar."

"Not to worry. The priest and I were speaking of a new and completely undeveloped fisheries when, without thinking, I mentioned your interest in the fishery and possibly importing wine at the same time, but no word of this will reach the vicar's ear. Father Niels will not break your trust. To ease your mind, step aboard and cast your eyes on the barrels of rosé we carry below decks. All nineteen of them are for you. We'll wait until after nightfall to load the kegs into this fine new storehouse. Only we will know."

"Nineteen barrels, you say? The small amounts our fishermen have been able to bootleg in for me have pleased many a Kerryman over at Bessie's this spring. They're forever pestering her for more and I like it too. Nineteen barrels should more than keep them happy, not to mention the commitment I made to the Sisters in Limerick. But tell me, what is this new source of fish you have become aware of?"

De Portu breathed a sigh of relief. "Well, since we are back to being friends, I will say we are going fishing in a special place, one never accessed by us Basques. If you will accept this wine in return for an equal number of kegs of pickling salt, on our return, those same kegs will be filled with all the cured fish your warehouse can handle."

"Your offer sounds much the same as the deal I made with a boatload of unpleasant Vikings. But I can't deny the value of the wine you have arrived with today. I will give you the salt you need, and upon your return, I'll be grateful to receive as much fish as I have room for. It has been near empty for too long. So tell me, where will you go to find the cod I need?"

De Portu was guarded. "Ah, I'm afraid it must remain a Basque secret for now, but I am confident we will both be pleased."

"Well, I thought this was a partnership. You can share *my* secret with a priest of all people, but you can't tell me *your* secret because you are a Basque."

* * *

Vicar Maurice's traditional Easter feast for his staff was dragging on longer than he had hoped. He expected several of his more favored guests would arrive at any moment, and he feared the servants would not have finished their holiday meal. Too fidgety to sit still, he told them to eat up and left the hall to climb up to the parapets, hoping to see his guests in time to waylay them.

What he saw in the distance was not his cronies approaching, but a tall man in full priest's attire, walking beside an ox cart. Maurice hurried down to welcome the unexpected visitor. They exchanged formal greetings in Latin, then chatted in French until Maurice spied the oak cask on the cart.

"It is a measure of a Basque wine," Nils explained. "The captain of the fishing vessel I arrived on today wishes to donate it for the local parish's use."

The vicar was delighted with the offering and had it sent down to the root cellar to be reserved for his favored guests after the servants departed. He invited Niels into the kitchen and offered him a plate of leftovers from their feast.

"So, pray thee, Father, tell me of your connection with foreign fishing vessels and more importantly, ones from wine-producing countries?"

"While in Spain, my duties included welcoming newcomers trekking to the shrine of Santiago de Compostela. You might know it as The Way of Saint James. I counseled the pilgrims on the privations and rewards of their journey. Until my language skills improved, I struggled to make myself understood, but the people there were patient, including the master of the vessel who offered to bring me here." Niels was careful not to mention the nineteen barrels to be stashed out of sight in Aideen's rafters. "The gift of this one cask was the captain's idea."

Maurice paced the length of the kitchen floor. "I'm curious. Does this sea captain often make voyages to our coast? If so, might he be encouraged to deliver more such gifts?"

"Many fishermen make such voyages. I am amazed at all the places they've been and how far they go. Indeed, the captain of *Espiritu* plans voyages well beyond our coastline and will return with enough fish to serve us all on meat-free days. You should know there are enough fish in —."

"More to the point," Maurice interrupted, "do many such vessels venture near to us? And if so, could a select group of devout Catholics from here make the voyage in reverse to Santiago de Compostela?"

Niels was delighted to think he could help establish a faith connection between Ireland and the famous shrine. "As I have traveled here from the cathedral, Reverend Father, your parishioners could travel back. Obviously, the fishermen have their own operations to consider first, but when their season is over, they could offer passage to many of your local crofters. If I assisted with such a pilgrimage, I would begin with advice on how to —."

Maurice held up his hand. "And these fishermen carrying passengers from here, could they each bring back a full cask of wine, as you have?"

"Well, having the vessel return with wine is certainly possible, but how will they deal with your import tax?"

"What import tax?" Maurice sounded offended. "I have yet to collect any tax at all. Last year, there was a tax suggested for foreign wines, but the matter was cleared up by our Vicar General. Perhaps I forgot to tell anybody, *mea culpa*, but let me ask again, could these sea captains carry passengers who will each bring back a full cask of wine, as you have?"

"I am sure any of your crofters who become pilgrims will agree, since it is for the church. However, I must tell you The Way of Saint James is arduous. The pilgrimage first demands a strong commitment and motivation beyond any mere personal interest in wine."

"Indeed Father, you give me ideas to ponder, but you must be eager to be on your way. Where did you say you are headed?"

"I didn't say, but my home is in Aarhus, Denmark. My bishop wanted me to confirm the existence of Vinland's great fishery and report back, but I have been absent for more than a year now and ...."

"I'll ask my page to find a mat for you in the staff quarters. He'll speak with the cook and have some pottage prepared for you in the morning so you can continue your travels. Before you do go, however, please leave me the name of this seagoing voyager whom you say might bring us wine. I would like to have a discussion with the fine captain."

* * *

With Niels accommodated, Maurice returned to gather his cronies for their traditional Good Friday banquet. After full servings of mutton, poultry, pork, and venison, he called for the cask to be brought up. Pulling the bung himself, he said, "Tonight you have been enjoying our local produce, but there is more. I am pleased to share this variety of rosé from somewhere in Spain as my preferred indulgence." He surveyed the faces of his appreciative audience. "And I am pleased to announce this will be the first of many."

Maurice's cronies raised their glasses as one in salute. "You have a contact with exporters of this fine wine from Spain?" one asked.

"I have a connection." The vicar stood and paced the room. "A Danish priest is sleeping below stairs who will set up annual visits of our peasants to the revered Santiago shrine in Spain. He speaks a dialect of Spanish and has close associates at the shrine. Best of all, he will be our connection to the wine makers there."

Maurice's mind was racing. "We will recommend that our peasants make one such voyage in their lifetime. And each one will be tasked with the return of a cask of wine or sherry for my ... for the needs of the parish."

Long into the night, the benefits of his new idea swirled in his mind. At daybreak, Maurice called Niels up to his chambers. "After you left us, our parish deacons decided upon a wonderful service you could offer for our community. You must relocate to our parish on a permanent basis and organize an annual fall pilgrimage along The Way of Saint James. You will arrange things with the captain of the high seas vessel, and I'll send instructions to your bishop in Aarhus to release you from his assignment. A copy will be sent to Rome. As a member of the parish staff, you will be given lodgings on the estate. After all, a pilgrimage is more important for you to occupy yourself with than fish, correct?"

"I am pleased to be of service, but there is one boon I might request of you. All fish should be tax-exempt as wine now is. After all, it is such an important part of Lent, and we need the extra fish landings."

"Yes, yes, such a trifle. Don't bother me with this issue of taxes again."

"You must excuse me. I'll rush to inform the captain of your decision before he departs."

* * *

De Portu stood casually watching his crew carry kegs of salt, baskets of biscuits, sacks of Dingle beans and straw aboard *Espiritu*. He was feeling discouraged after the conversation with Aideen had ended on a sour note and couldn't dismiss the error he had made. *I should not have played such a silly trick on her. I came here to be her partner, but all I said was the location 'remains a Basque secret'. How could she trust me now?'*

The line of supplies continued to be loaded until the cabin boy began struggling to wrangle a pair of Galway sheep up the gangway. The crew stopped working and made fun of him until de Portu became angry. "Get the provisions below, except for the sheaves of straw. Pack them in the dory," he barked.

Elkano, the Second Mate, was the first to question de Portu's abrupt directive. "We do need straw, Captain, but more important would be three or four dories for when we get to the fishing grounds. If we have but one, and it remains full of straw, how are we going to overhaul our herring nets?"

"You should be pleased. The straw is extra bedding," de Portu explained. "We will be underway for longer than usual, and your mattress bags will need restuffing. There'll be a change after twenty days, each one on a rotation. The used stuffing, full of lice and too compressed to be of any value, will be layered under the sheep and then dumped overboard after they use it. As to the dory, I'll explain after we are underway."

He called the cabin boy over. "Hang some slings in the stall below decks to suspend the woolly beasts from the deckhead so they don't fracture their spindly legs. And I know it's not nice, but the crewmen all have bunks down there, so you must do a thorough job of mucking out the stall each morning to keep the stink down." No happier than the animals, the boy groaned and hustled them down the hatch.

The captain's instructions left Elkano agitated. "Captain, you seem in a foul mood for some unknown reason, but there is a more serious issue. How far will we be sailing? If we get a change of bedding straw every twenty days, where are you planning to take us?"

De Portu was ready to lay his plans before the crew when Aideen walked up the gangway. She swooshed one hand in front of her nose. "I can't imagine how those beasts will smell after a few weeks at sea, Captain." Holding a basket of salted fish in the other, she said, "This will keep even longer and still smell better. When you want to eat it, just soak it overnight, and it'll taste fresh-caught. It's my way of apologizing for being short-tempered with you." She wished them a safe passage. "*Slán libh*. Safe travels."

For de Portu, there could not have been a better send-off. He wanted to embrace her, but she had already turned to step back down to the dock.

* * *

The vessel slipped into the current when Niels rushed up to the wharf and bellowed at the vessel.

"Captain, I've hurried over with the latest news. Two of the best things have happened. The vicar has decided there is to be no more import tax on your cod."

"This is welcome news, Father," De Portu shouted back. "You and Aideen must work out the fine points before my return. If all goes to plan, we should meet again two months from now."

Aideen turned to Niels, relief bringing a tear to her eyes. "There is no more tax on imported cod? Well, if de Portu is successful on this voyage, his deliveries will be my salvation."

Niels wanted to keep shouting to de Portu, but knew the vessel was beyond hearing. Bursting with his other news, he said, "And there is more, Aideen. Your vicar will announce today that his tax on imported wines has been revoked. You know, I got the feeling his tax was blocked by powers higher than him some time ago, but he avoided telling you."

Tears welled in the corner of her eyes and trailed down her cheeks. "I don't know if I'm more upset at the vicar for deceiving me or at myself for the unfriendly welcome I gave you yesterday. You have been here less than a day and already worked two holy miracles." Aideen hugged the priest she had dismissed so impolitely one day before.

"And speaking of things holy, it was the Reverend Mother in Limerick who spoke to the Vicar General regarding the tax for me. I must send her a good part of this wine as thanks."

She set off for Bessie's, hoping Paddy would be there and willing to make another trip to the nunnery.

* * *

"How would it be if I brought you back some salmon Aideen? There has been a good run this spring. Mind you, it's all but done now, but if you might not have enough fish to fill your storehouse and you're willing to starve yourself, a few salmon are always appreciated by those in the know. I'll fetch you back a feed."

"Thanks, Paddy. That gives me an idea. After you're done in Limerick with the nunnery's wine, it would be good if you could find enough salmon for Lil to put on a grand feed. I owe a big favor to her for all the provisions she let me cart off this spring when I was building my storehouse. And who appreciates a fresh salmon more than a cook? She could collect the manor staff for a traditional feast on me. We'll make it a true

Easter banquet, far better, I'm sure, than the staff got from Vicar Maurice."

* * *

Five days later, Aideen led her mule cart back to the root cellar. She pried it open and carried four baskets of large, fresh salmon into the darkened space. Lil was quick to drop what she was doing in the kitchen and trot down to see who had entered.

Seeing the baskets of large pink fish, Lil said, "Now there's a treat for each one of us here. After you told me you might be able to provide the salmon, I mentioned it to a few of them, and they all agreed. 'That is so like our friend Aideen to do something for us after what the vicar did to her,' they said. They miss you, my dear, and it will be splendid to have you join us. I'll pick a day next week when the work is done and, more importantly, a day when the vicar is away on his pastoral visits."

"And I'll bring some of my new batch of Basque wine," Aideen added. "It's not to everybody's liking but it comes to me without the vicar's tax so I'm keen to share some around, if only to spite the man. We must celebrate the small triumphs, right?" Lil laughed and said she had something else on her mind.

"Sit down on these crates, dearie, I've something else to make you smile. I think you know how the vicar enjoys his rosewater? He talks of nothing but the good things it does."

"He drinks rosewater now?" Aideen asked.

"No, no. He bathes himself in it, then spreads it around with a special sprinkler he purchased from a traveler from the east. He thinks the fragrance makes him elegant."

Aideen thought the idea was quite peculiar. "I think the sweet blossoms smell heavenly, but I'm not sure I'd want any man of mine to smell of roses. Anyway, how do you make water from the petals?"

"He has me collect the petals regularly from the hundreds of bushes he planted. I must crush the petals in a mortar and steep them to make a paste. Do you know how many petals I need to boil to make a bottle of the stuff? He says it cures his pimples. Now, I too enjoy the smell of the wild Irish rose, but not so when there are hundreds of them at once and they're all rendered down. Not to mention, his plants have tiny thorns, making my fingers swollen. I'm a cook. How am I to prepare his food with stubs of fingers as sore as these?"

Aideen wanted to be sympathetic. "Maybe if a little of the scent sticks to him, it will soothe his black heart."

"Well, let's hope so, but here's the curious part. He has me prepare enough paste for his bath, too. He's after soaking in it for his hemorrhoids."

Aideen winked at her old friend. "Then perhaps you should leave a few of the tiny thorns hidden in the paste."

Chapter 25

At sea, off Dingle Harbor

When de Portu could no longer see the mist-covered hills behind Dingle Harbor, he called his crew. In a direct and confident manner, he outlined the plans for the trip.

"We won't be setting nets for herring. I don't have to tell you they are scarce out here, so we won't be wasting our effort."

Elkano spoke up first. "What do you mean 'wasting our effort'? We have to return with at least some herring to feed our families. Have we come here only so you could chat up your woman friend, Captain?" He waited, expecting a full explanation.

De Portu squared his shoulders. "We are going fishing, but we won't be hauling half-empty herring nets. I ordered new handlines for you all, including one for me. Dingle was our last stop before we visit a place teeming with fish. There'll be plenty to feed the folks at home, too much, perhaps. This is to be the greatest trip you could ever imagine." De Portu sensed no one was buying into his enthusiastic revelation.

Elkano stepped up again. "How far will we have to go? You have to tell us, Captain, since we'll be well beyond the herring grounds shortly."

"How far? The far side of the ocean is where we're going. It's called Vinland. I didn't tell you before because the exact location was unknown. In fact, it still is, but the Danish priest we brought with us visited Vinland himself last spring. He described huge schools of codfish there. The fishermen filled the vessel he was on in one day."

To prove it, de Portu rolled the priest's sketch map flat out on the deck and took a sheet of figures from his waistcoat. "You already know one-third of the price we receive goes to you. Here's how valuable I figure your third will be." He passed the sheet of calculations to Elkano, who reviewed the sheet and showed it around to the crew.

"Captain, these numbers are good, but what you're saying is you plan to cross into the unknown searching for uncharted land. No Basque fisherman has ever imagined such an undertaking. No fisherman from anywhere has. And say it is all true, the priest's sketch is nowhere near complete. It doesn't tell us how many days we'll need to get there, nor what the sea conditions might be."

"We Basques leave our families for months each season," de Portu continued. "Rough seas are no concern for us either. If you think of the earnings we'll share, I don't understand why you're not cheering."

"Yes, Captain, we do leave our families for months when we know where we're going and if there is a good chance we will come back with a full purse, but you haven't said what we will be fishing for. As far as your priest and his map go, we all know priests don't need to make a living. Is that to be our fate too?"

The crew was quietly edging over toward Elkano. De Portu could see that a distinct standoff was developing.

Shicki tried to ease the tension. "You're right, Elkano. Vinland will be a long voyage, several weeks or more. It's the reason we have the extra bedding. The best part is that loading this boat will take no more than a day or two. In less than a week the catch will be in the pickle and we'll be turned for home. You spent weeks on this vessel last season before we spotted one whale. How many more weeks did we spend dragging it to shore, cutting it up, and rendering the oil? In the same amount of time we need to hunt and render down the oil of one whale, you'll become richer than any whaler or herring fishermen ever was."

Shicki's argument had eased the men's fear, and de Portu read the change in mood on the faces of his crewmen. "Elkano, I know someday you want to have a vessel of your own. So, all the days and weeks we spend on board *Espiritu*, you can be thinking of how your share of the trip will pay for the new vessel."

"And the rest of us, Captain?" Several nodded their heads in agreement. De Portu didn't see who asked the question, but knew he had to do a better job of convincing them. "I have one other thing to show you." He crossed the deck to the locker under the forepeak and called for help, dragging the weighty sheet of rigging up to the mast. "This mainsail is lateen cut. It acts the same as the wings of the birds we see leagues out at sea."

"See how it resembles the shape of a wing. To beat into the wind, we keep the cloth at an angle between the wind's direction and our heading. This shape won't trap the air as our square rig does, it lifts our boat over the surface. We'll run true and let the keel keep us steady."

As soon as they hoisted it on the mast and secured the leading edge to the forepeak, the canvas went taut and the boat lurched ahead.

"We're flying," Elkano shouted. "We're flying *into* the wind."

Espiritu picked up speed, heeled over, and began carving a clean path through the chop. Even de Portu was surprised how fast the ship was going. He called out to the crew. "Shicki, set our bearing and have this canvas drawn tight."

Espiritu lurched again, still heading straight into the wind. The crew gasped.

"Take a last look back, lads. The next land you see will be a whole new world. You will get many questions when we return, but no word of this new rigging will be spoken by any of you. Lie through your teeth if people ask how fast we traveled. The answer is simple ... 'farther and faster than any Basque before'."

* * *

Two weeks out of Dingle, Shicki noticed the men still enjoyed how their vessel was taking advantage of the winds, never mentioning the extra work needed to keep it trim.

"Captain, we have been at sea without sight of land for longer than any of us has been before, yet not a one has mentioned it. The food locker remains well stocked, and the sleeping mats still have some spring in them. Nobody has complained." He pointed to Elkano, who was standing at the forepeak. "Do you see the grin on the Second Mate's face?"

"He's calculating how many leagues we have covered," de Portu said. "And I'd say he's impressed by the number."

"Did you look at the surface of the water, Captain?" Shicki asked. "For most of our way, it was a deep gray, like your pewter plates, but now we are splashing through green waves and the surface is brightening into a deep blue. The air feels different too. Do you think there's land here?" Shicki trailed a line with multiple hooks and a sinker over the stern. "If there's land here, there'll be fish here."

When he caught nothing, he groused, "Either there's no fish or we are moving faster than the fish here can swim."

* * *

The number of days at sea had stretched to twenty-one. There had been no sign of land, nor shorebirds, nor anything to give him assurance of seeing land anytime soon. De Portu paced the deck, wondering what he was going to say if the situation was the same after another seven days. To hide his nervousness from the crew, he retreated to his cabin and reread the priest's notes. The same vexing questions kept repeating in his mind. *If we don't land at Vinland, where will we end up? How much farther can we go before it is too late to turn around?* After four hours, he chastised himself for hiding away and stepped back out on deck. He checked over the stern at the vessel's wake. It was reverting to a sullen gray. *Why didn't the priest speak of the change in the color?*

When darkness fell on the twenty-fourth night and the clouds parted, de Portu sighted the Pole Star. After he took a bearing and reread the notes Niels had given him, he conferred with Shicki. "I've checked this astrolabe when you mentioned the color change of the seawater, and again now. We are not on the same latitude. It's as if we drifted north, but now we are being carried back south, and the color is different again. I have no explanation for these changes. The priest didn't mention the olive shade of the water we crossed through last week. I think it was because the Vikings came down from Greenland to Vinland. Coming from there, they didn't experience this same current we're fighting."

He and Shicki withdrew into his cabin. "If my calculations are anywhere near correct, we should have sighted land long before this. We've covered twice the number of leagues I predicted we would."

"So, your calculations have not been correct, Captain. You saw the mysterious green current we sailed over. Maybe it did push us north, but now the current is gone, and we are slipping back to the south. Do you think we might have passed by the fabled Vinland when the current forced us north? Could it be behind us?"

De Portu couldn't dare think Shicki was correct, but he did record his observations in the logbook.

7 June. Twenty-five days at sea.
Strong currents forced us north but then died away.
Now moving with a cool current to the south.
Doubts growing among the crew. Have we sailed right past Vinland?

Chapter 26

Notre Dame Bay, Vinland

At dawn, low clouds blanketed *Espiritu*. Dew dripped from the lateen rig, which hung limp, fluttering aimlessly against the mast. De Portu stood at the helm, struggling to make out any sign of land and remembering how a crew's spirit sinks when they are becalmed.

"Shicki, this cloak of fog feels the same as the low clouds that often cling to the nearshore fishing grounds at home." A pair of black birds scampered across the calm surface in an attempt to get airborne. "And those are cormorants. At home, they're never far from shore. It's possible a landfall might be waiting for us on the other side of this fog bank."

"A HORSE'S HEAD!" The shriek from the lookout echoed across the surface as he bellowed from the crow's nest and pointed to a huge snout fringed with long whiskers that stared up at them. The deck took a sharp list as each man on board rushed to the side rail. "And look there, there's more farther out. This place is infested with them."

"Get back to your stations," barked de Portu. "It's no horse, but it is the largest *seal*, with the biggest black eyes I've ever seen.

And men, what do seals eat a lot of? Where there are seals, there are fish. And another thing tells me we must be close, Shicki and I spotted two shorebirds."

The crew strained to see through the mist, hoping to be the first to sight the land they had been promised. Shicki uncoiled his handline again, tied a strand of red yarn to the hook and flipped it over to see if his captain's prediction was correct. He felt an immediate tug, followed by another.

Everybody on board, including de Portu, began handlining, splitting, and salting an unending stream of the most beautiful fish they could imagine. The cloud cover lifted at midday, but nobody, not even the lookout, noticed the low rocky islands dotting the surface of the sea a mere league off their port bow.

When Shicki landed the heaviest cod yet that day, he struggled with both hands to lift it above his head and show how it was as long as he was tall. "It's a whale," he yelled. As he lifted his head, his gaze met a tree-covered cliff rising skyward. He dropped his trophy and began shouting orders.

"Lower the canvas. Watch for rocks before one scrapes the keel. Ready the sounding lead. Stow your gear. We have to maneuver out of these shallows." Jumping over fish heads, hooks and each other, the crew hurried to prevent their vessel from drifting any closer to the cliff face.

* * *

For de Portu, his long-sought objective was in front of him, but the excitement of sighting land led to so many questions. Will the afternoon air be too hot on the beach rocks and cook the fish before it dries? Will handlines be the best way to get all this fish aboard? How soon will the holds be filled? He stepped back into his cabin to record their position in the logbook.

*8 June – Fishing in the largest school of cod we have ever seen.
The priest was correct. Estimate the latitude to be fifty degrees north.*

He stepped back on deck as *Espiritu* slipped past the headland and into a wide bay. The eastern side was a scenic ribbon of rocky shoals and verdant, tree-covered islands. Further in, they spotted what appeared to be the mainland and a crescent-shaped cove.

"There's a cove over there. Take her in closer, Shicki. We'll send in a shore patrol. You, Elkano and the two strongest men should go."

As the four crewmen clambered over the side and down to the dory, Elkano voiced the serious concern he knew the whole crew was thinking of. "What if there are people already here, like the *skræling*?"

"I'll be honest with you," de Portu admitted. "I have no idea what to expect. This is their land, so I don't know how they might react. That's the reason I'm choosing you four. I know we all want to set foot on a new land and see what's beyond the tree line, but you four are the most able to handle yourselves if there is a confrontation. Locate a stream, take a quick walk around and report back."

The rest of the crew, covered from cap to boot in fish scales, lined the rail watching until the little boat ground out on the shore. De Portu summed up the scene to anybody who could hear, "You're hooking so many fish, they are flying over the rail. They're all around us, underneath, too. We should have baskets to haul them up."

After an hour, the dory bumped up against *Espiritu*. Shicki scurried up the ladder. "Captain, you should see the young salmon in the brook. We found a great place for drying racks, with a southeast exposure, too. It's flat and covered with small rocks the size of the cobbles we use for paving the streets. The drying racks can be set right there, so no fish rests on the rocks when the afternoon sun gets too hot. There are berries on all the bushes, and on top of it all, a waterfall with the sweetest water tumbles down out of the rocks and runs down to the beach so we can rinse the fish. And you know what else a brook is for, to wash the stink off all of the crew."

"Was there a sign of anybody living here?" The question broke Shicki's enthusiasm.

"We didn't see any signs at all, no tracks, no shelters. I'm sure there's nobody here. Who could live in those thick woods, anyway?"

De Portu had an answer. "We'll find out soon enough. Father Niels told me how the Forest People were very wary. 'They can appear without any warning', he said. If and when it happens here, I'm hoping it will be as trouble-free as the priest described."

De Portu turned his attention to the freshwater consumption on the voyage so far. He didn't begrudge the amount each man was consuming, but he needed to know how many full barrels they would need for their return voyage.

"On the way here, we emptied four of our seven puncheons, so we'll refill them. Shicki, take a pair of the empty barrels into the waterfall you saw. Take in two men with carpentry tools to get going on the drying frames. With all these fish coming aboard, we can't wait any longer."

With fish coming in and men going ashore to set up operations there, de Portu began to relax. *Time for a little fun*, he decided.

"Elkano," he called out. "Have the men tip the three remaining full barrels over the rail to empty them. We should refill one as a spare, just in case, but we could use those two remaining barrels for fish. Or we could flip all the livers into them. They'll be rendered down to a tasty oil before we return."

Elkano balked. "You'll ruin the barrels."

"But we'll drink the fish oil. It's good for us. For any leftover oil, well, we'll sell it to the lighthouse keeper at home. I know it only fetches half the price of whale oil, and it burns smoky, but he'll accept anything for burning in his light."

Elkano didn't bother reacting, but watched the captain, just in case he did fill the barrels with livers.

* * *

"CAPTAIN."

It was Shicki's voice but de Portu couldn't see him. He peered over the rail and over to the far shore. There, heading for *Espiritu*, his First Mate was rowing as hard as he could. When he reached the vessel, he grabbed the rope ladder and scrabbled up on the deck, too agitated to speak.

"They, they have the lads. They came out of the forest and took them back with them."

"Who's *they*? Who took them?"

"The Forest People. I was loading the water barrels into the dory. I got one in and when I turned back for the other, all I saw was a pointed spear. They came up behind us. The lads had peeled the trees for the drying frames just as the Forest People forced them back into the woods. I made it into the boat, but they seized the others."

"What happened to our tools, the axes, adzes and hammers?"

"I didn't stop to check, only rowed like there was a fire in my arse."

De Portu frowned. "We *need* those tools."

"We need our *lads* more."

"Yes. Yes, we do." De Portu was embarrassed. He dropped his gaze to the deck, struck by how the situation had changed in an instant. He had no plans for what to do when they met any locals. A few seconds of silence gave him a chance to clear his head. He pointed to the two men who had gone ashore on the first scouting trip. "You two fetch the fresh water up on deck and put a tub of fresh split fish in the dory, not the salted ones. Afterwards, fetch me the ewe and come with Shicki and me. I'm expecting they've never seen such a fine beast. We might be able to trade her for our men."

Shicki was still in shock. "What if they grab the animal and eat her?"

"Yes, such a thing might happen, but we'll keep the male on board to save for barter later. The Forest People will understand their value as a breeding pair."

***

The dory ground into the cobble shore, and each man was sure he could feel pairs of eyes watching from within the trees. When they stepped out into the surf there was no sign of the tools, nor the hewn timbers. The ewe bawled and leaped from the dory, its hooves sliding on the wet rocks. De Portu grabbed her tether and wondered if the commotion might generate a sign of activity from within the forest.

"Look closely. There must be a path or an opening in the brush," de Portu instructed, keeping a tight hold on the animal.

It wasn't long before they could make out a faint line of footprints leading into a denser part of the understory. Not a word passed between any of them. The footprints became a well-trod path opening into a quiet meadow. The air had blown constantly at the shore, but now it was still. No birdsong could be heard, no animal wandered through site.

Three oblong shelters covered in animal skins and birch bark sat at the far side of a clearing, next to a babbling brook. "There's the reason why our first team didn't spot any structures. They are the same shades as the trees," de Portu said to nobody in particular.

A central fire pit smoldered with two plucked geese roasting on a spit. Another was lying on the ground until a place was available. De Portu called out the names of his missing men. A child darted out of the first lodge and headed for the men until a female dressed in a long flowing robe of animal hides appeared. She scooped up her child and ducked back inside.

"There must be more people in the lodge. Poke the ewe," de Portu suggested. "Let's see what they do when they hear her. Make her bawl."

The ewe did bawl, loudly, shattering the silence of the meadow. It worked. Three figures emerged, one woman and two men. De Portu and Shicki took a tentative step forward. Each group stared at the other.

"Show no anger," whispered de Portu, "but no fear either. They might be as surprised by our arrival as we are to find them here."

With the tether in one hand, de Portu extended his open hand in front of him while continuing to walk closer to the group of three. Shicki matched de Portu's steps, a pace behind.

The woman began speaking in a steady tone. Her words sounded harsh, but familiar to de Portu, but the expression on her face offered no clue as to her meaning. "Is she trying to speak Norse?" he asked Shicki, not expecting an answer.

He released the ewe, hoping it would wander over to the woman. "It's our gift to you," he said in Norse, with as much sincerity as he could muster.

She showed no sign of understanding him, but the sheep caught her eye when it lowered its head to munch on the fresh meadow grasses.

The woman spoke softly over her shoulder to two men moving nervously in behind her. De Portu whispered to Shicki, "She doesn't seem threatened by us. She must be telling them to relax."

De Portu tried to scurry the animal along with another kick to its rump. It jumped three steps forward, and de Portu signaled to Shicki to move a few paces closer with him. "She's a gift in exchange for our crewmen," he said.

The woman bent down to touch the sheep. She rubbed its soft wool with her right hand and gestured to the men behind her with her left. They disappeared back into the lodge. The animal continued to grind a mouthful of grass while the woman took firm hold of its tether.

Two frightened but relieved carpenters tumbled out of the lodge and darted straight for their captain. De Portu tried not to show any reaction, but standing little more than an arm's length from her now, he expected the woman would read the relief on his face.

"*Beothuk.*" She pronounced the word as a non-committal assertion, extending both her arms wide.

De Portu relaxed and tapped his hand on his chest. "Gabriel de Portu."

A handful of villagers flooded out of the other two lodges and stood, regarding the strangers with a mixture of fascination and caution.

"What do we do now?" Shicki asked.

"Well, nobody has any weapons, so act as if this is nothing unusual. We're here for the fish, and we need their cooperation to get it."

"And to dry the fish, we need our tools back," Shicki added. "The racks won't build themselves and it must be done soon. All the tubs onboard are overflowing already."

"They don't need to know how much we need the tools. Follow my lead here. Take the two men and go back to the dory. Fetch the tub of fresh fish."

The woman spoke again, sounding more earnest. Again, de Portu couldn't tell from her expression what she was thinking.

It was a standoff until the box of fresh, trimmed fish was carried into the clearing and set at the woman's feet. She stooped and smelled it, admiring how it was cleanly split and deboned. Her blank expression gave way to a slight smile.

"*Bobboosoret*," she proclaimed.

"We hope this will be enough for your village tonight. We must have our tools back in return, however." De Portu was trying to sound as polite as possible, while making gestures he hoped would be interpreted as using hand implements.

She spoke to the man next to her. The hammer, axes and adzes appeared within minutes.

De Portu offered a half smile. "I guess you plan to keep the poles the men cut, as long as we can cut a few more trees from your land." He told Shicki to send the carpenters back and to keep them busy until they completed the drying racks. "And tell the men to ferry in all the fish piled on our deck. They'll need to rinse it and spread it out to dry."

The woman stood staring until, as if she remembered an urgent problem, she turned back to the lodge. Sensing their exchange was over, de Portu went back to the shore to check on his two traumatized carpenters, hoping they had calmed down enough to go back to work.

"The ordeal must have been frightening for you," he offered. "You did the proper thing, though. You didn't panic. What a tale you're going to be able to spin back on board."

De Portu and the two carpenters left the rest of the men working on shore and rode back on the dory's return shuttle. As soon as they climbed back aboard, the crew peppered the two crewmen with questions. De Portu interrupted, "Remember how much our next few days onsite will be worth. We won't fill *Espiritu* with fish without their help. We have much to gain with this fish. Let's not cause any problems. We have to be more than careful to not start a row."

During the night, other problems filled de Portu's head and kept him from sleep. *We've already caught more than Aideen can handle, more than she ever expected we would deliver, so will the balance of the load find a sale at home?* The ominous question stuck in his mind until dawn.

***

On their third day, the coast was again obscured in low clouds. As the haze lifted, glistening schools of cod could be seen circling below the vessel. De Portu was tempted to continue fishing, but *Espiritu* was already at its limited capacity. Standing on deck, recording the numbers of tubs going ashore and inspecting the returning fish being packed aboard, he kept a careful watch on the vessel's stability.

"I want to visit their encampment once more. We'll be sending the ram in too," he announced to nobody in particular. "He stinks now, so we'll all be glad to be rid of him. Once his pen is scrubbed, at least we'll have his space for more fish."

The captain spoke directly to the animal. "I hope your woolly mate isn't rotating over the fire pit with those two geese. I plan to offer you as a way for them to start their own flock."

With the skittish ram secured in the dory, de Portu and Shicki rowed for shore. When they landed, they could see a heated scrum developing near the drying racks, and de Portu immediately stepped into the middle.

"They tried to eat our fish," one crewman complained. "And I guess they don't agree with what we're doing to it, but it's our fish, damn it."

"It's not our fish yet," de Portu insisted. "We need to keep both sides on the same side until it's dried and stowed aboard."

He noticed the local man who tried to eat the salted fish was still spitting the bits out.

"It tastes odd, eh? I'm not surprised. I guess after we gave you the box of fresh fish, you expected you could help yourselves. What we do by salting the fish doesn't make any sense to you, but I'm glad to see you think it tastes bad."

He handed the man the sheep's tether. "Here, take him. Its offspring will taste much better than our fish. Shicki, bring in more of the latest fresh fish for them. We have more than enough for them, too. Have the crewmen who still want to visit the shore brought in to gather beech leaves and bundle them back to *Espiritu*. We'll be needing them as bedding."

Shicki waggled a finger and pointed into the woods. "Those aren't beech trees, Captain. I don't think this land has trees to match the beech at home."

"Well, have them gather sacks of any fresh broadleaves they see and hustle them back aboard. Tell them to explain that these ones are better than the beech leaves at home."

De Portu gave a nod to the man still holding the animal's tether and indicated he wanted to go back with him to the lodges. As soon as the group broke into the clearing, the ram detected the ewe's scent and began bleating. A similar response came from behind the group of lodges. De Portu stroked the ram.

"At least your partner hasn't become lamb stew. Let's go coax her to come out."

Instead of the ewe, a tall lad with a shock of blond curls emerged from one of the lodges. He too was dressed in animal hide, and de Portu took him for a local until the man greeted them in perfect Norse, "You men must be off the caravel sitting at anchor in the bay. Am I glad to see you!"

"Yes," De Portu gasped. "I'm Gabriel de Portu. But who are *you*?"

Chapter 27

Notre Dame Bay, Vinland

Benji opened his arms wide. "My name is Benji Leifsson. I'm a Greenlander, but I was on a Norse longboat fishing in the bay earlier this spring. The Viking captain and his brother, the cowards — they left me here. They told me to find my own way back to Greenland. I'm going to find them someday and when I —."

"You're the second person we've met who had trouble on a longboat. I bet it was the same Viking brothers. Blessed Mary, how can they be so cruel to their own kind? They're worse than foxes in a hennery."

Aideen's story of the longboat crew disappearing with a load of her salt came to de Portu's mind too. *Could they be the same crew, too?* de Portu thought.

Benji continued bubbling over with joy at being rescued. "After you showed up two days ago, a runner was sent to the camp where I've been living. They took me in. Small villages are scattered all around this bay."

"What you're telling me explains why the first woman we met here was trying to speak Norse to us. I had no idea what she was trying to say. Maybe she took us for Greenlanders."

Benji nodded, "I tried to teach them a few of our expressions, but they think my Norse language is too complicated so I ended up learning theirs."

I think your Norse words sound as if they are all said backward, and their language sounds even more confusing than yours."

Benji kept chattering. "One woman made me feel welcome and taught me enough to get along. She taught me personal things too, things men don't share, ones my grandparents never told me. Did you know women —."

De Portu tried to calm Benji down. "Tell me, do you know what *'Bobboosoret'* is? I think it was the word the woman used. Is it her name?"

"Ha ha. No. They call themselves Beothuk. *Bobboosoret* is their word for codfish. She must have noticed how you split the fish open and was reminded of how I did the same thing. They think it's a good idea, but can't understand why you use salt and then keep it stored away. For them, whenever they get hungry it's too easy to find fresh fish."

"*Easy to find* is no exaggeration in anybody's language. And we'll be back for more. But right now we're near ready to leave for home. Maybe you want to come back with us?"

Benji agreed instantly. He turned back to the lodge to say his goodbyes.

* * *

De Portu stood on deck, keeping a close count of the incoming fish and directing where they should be stowed.

"Fit 'em in tight, men. The load has to be rock solid to keep us on a steady keel when we get on the open sea. Fit as many as possible in my own cabin, too."

By nightfall, Benji joined the dory ferrying the last tubs coming aboard from the drying racks. The dory was secured aboard, and it too was packed full with fish. They left all the tools on the beach, a parting gift.

"We'll be heading north and east to Greenland, mostly running with the winds at our back, Captain," Shicki stated. "So, do you want the other main sail brought on deck?"

"Yes, *Espiritu* will hold her trim better, and it'll be less work for the crew, but remember," de Portu took a defiant stance. "The lateen sail and our location here are confidences you will all keep. Nobody should ever be told how we've managed to fly this far over the sea and where we ended up."

De Portu's stern look was replaced with a grin. "There's more news. You're getting fresh stuffing for the way home. They say our King Alfonso will sleep on nothing but a linen bag stuffed with beech leaves, and we found enough leaves for you all to restuff your mats. And they are beech leaves, right, Shicki?"

"Yes, springy green leaves from right near the beach, they are, Captain."

De Portu had bad news for the crew, too. To make room for all the fish, the sleeping space is now limited, and you're going to 'hot-bag'. I mean if you are on watch, you'll be on deck. If not, you'll be sleeping on the leafy mattress bag of another crewman who is on watch."

* * *

Moving away from their mooring a stiff sou'wester came up. The breeze blew away the smell of cured fish and chilled the crew. De Portu called the men in a circle again.

"After we drop this unfortunate Greenlander off, we'll turn due east. I can't predict how long it will be before we make landfall, but as long as we keep our cargo stable, there'll be an eager buyer waiting wherever we tie up. We'll unload as soon as we land. Afterwards, we'll ride our empty vessel home to San Sebastián. Imagine how we'll toast our fortune come evening at *Bar La Navarra* for the whole winter." Each man stopped to do a reckoning—their cargo meant the finest payday they ever imagined. A cheer rose from the deck.

Benji tapped de Portu's shoulder. "You know, I don't want to go back to Greenland right now. After being so long away, my grandparents will be at peace with me being gone. At best they'll think I found a girl in Ireland or worse, maybe lost at sea, but how can I show up now only to tell them I plan to leave again, as soon as another vessel comes along? Am I welcome to stay as part of your crew, wherever you're going?"

De Portu was cautious, thinking as he was talking. "Well, it would save us many sea leagues. If we head back to Dingle on the coast of Ireland, we'll stop at my main fish buyer. We could drop you with her or you can come with us to San Sebastián, whatever you wish."

"Dingle, you say." Benji's voice cracked. "Heading to Dingle would be tremendous for me because I met the lady there who buys fish there. I've been hoping to see her again." Not only did he want to see Aideen, but in his mind, he always hoped he could find out where Arne and the crew ended up. He might be able to get more information there.

De Portu nodded and called his mate. "Shicki, we have new plans. We're heading for Ireland. Fetch me the logbook. I should be keeping a record of this."

After recording the date and position, de Portu checked his compass and took a sighting with his astrolabe. He penned one last entry.

Departed from estimated forty-six degrees of north latitude with the fullest load ever. Heading east northeast with favorable winds. Estimate eighteen days to San Sebastián, by way of Dingle, Ireland.

* * *

Benji was watching him marking their position. "Forty-six north is not correct, Captain. I remember Sweyn, the navigator the Vikings picked up in Bergen, he figured this was fifty degrees."

De Portu gave him a cheerful smile and tucked the logbook into his vest. "If this entry is ever read by another captain, he might head for wherever forty-six degrees north is, but he won't be in our Notre Dame Bay, and he won't find our codfish.

"I don't understand, Captain. The fishermen at home say nobody could ever catch all the fish around here. They also say more fish live off in the deeps. For sure, other fishermen will stumble onto them before long."

De Portu was aware he couldn't keep a lid on the news of such wealth for long, but defended himself saying, "We are Basques. Yes, another captain may well arrive at this amazing place someday, but whoever he is he will know I was here first, Captain Gabriel de Portu and the crew of *Espiritu*."

The next morning, he called Elkano and Shicki into his cabin. "Nobody at home will be wise to the best fishery in the world if we first unload a share of this at Aideen's. In fact, we should unload as much as she will take. We will fill her space right up to her rafters so there'll be less need to find buyers for at home."

Elkano reminded him that the men looked forward to some bragging. "This will be the largest load ever landed in San Sebastián."

De Portu's answer was straightforward. "They won't brag for long if nobody is interested in salted cod. And remember, we agreed to sell to Aideen first. Perhaps we won't earn the reputation as the best fishermen San Sebastián has ever seen, but we will be the richest."

* * *

Their square sail ballooned out, and a noisy wind made it hard to hear each other on deck, but several crewmen heard a commotion of rumbling harrumphs and whistle sounds on the far horizon. The noise increased until the lookout spied whale blows less than three leagues in front of them. Moments later, he shouted, "There. It's the tail of a Black Fish, what the

Spaniards call our Right Whale." He scanned the surface and spotted two more lumbering black forms breaching, each longer than *Espiritu*.

Elkano vibrated with excitement. As much as he was pleased with the load of fish they carried, seeing so many whales in one place thrilled him beyond belief. Surveying the huge pod, he shouted, "There's a hundred of them, maybe a thousand. They're all blowing and turning tail. This is where all those huge black creatures disappear in the summer. They must follow the cool currents, too." He threw up his arms at them. "You are lucky, my friends. You must have heard how our holds are already full, and we can't be making room for you today."

Elkano had no doubt what he would do with his share. "Captain, I'll be getting my own whaler next spring, and if your man in Hendaye will fit me out with a sail, I'll show these whales how lucky they are, or not."

Chapter 28

Dingle, Ireland

A tall ship appeared on the horizon, its square rigging backlit by the setting sun. "The vessel out there is tall enough to be *Espiritu*," Aideen shouted into the air. "And it grows larger as I watch. Even better, she is *Espiritu*, exactly what I've waited for these two months."

She threw her arms up in the air and waved as high as she could, hoping it would indeed be *Espiritu*, then dashed back inside to tie up her tresses and shoo the swallows from her near-empty storehouse.

Espiritu tacked and jibbed its way into the harbor entrance and at last Aideen could make out de Portu standing at the helm. When she saw who was standing next to him, shivers ran down her spine. "From a distance, he could be Benji." The words came out louder than she wanted.

When the vessel was within hailing distance, she could clearly see it was Benji. The shivers became a knot in the pit of her stomach. She called out, "Captain. You surprised me when you tied up this spring with a priest in tow. Now you show up here with a tall blond Norseman, Benji, my special friend who has been gone for a year. You are a true savior of lost souls."

The gangway swung down and bounced with a thud as it hit the wharf timbers. Benji jumped down to greet her and wrapped her in a hug. She felt his hot breath at the nape of her neck and sensed his lips brushing against the bare skin there. In that moment everything in her world stopped, except for the tingling down her spine.

"So many times I hoped I would see you again," she whispered.

He loosened his hug and looked at her. "Last year when Arne and Stok left your dock, it was the beginning of their most outrageous scheme yet. After the winter in Stavanger, we left for Bergen to steal supplies from the guild there and crossed back over to Vinland. But as soon as we loaded the longboat, Arne and Stok had the crew abandon me there. I was marooned."

Memories of Arne and Stok flashed into Aideen's mind. She regretted ever thinking she could deal with them. "They abandoned you? By God's bones, they are cold-blooded." Aideen dropped her arms and stared at him, bewildered. "They marooned you. Did you see any Forest People?"

"Yes, the Beothuks," Benji corrected her. "They took me in. I've been living with them."

Aideen dabbed at a tear sliding down her cheek, but noticed de Portu stood watching them. She wasn't ready to show de Portu it was a tear of joy for seeing Benji.

"See, Captain. You bring me to tears. This tear is because the scent of salt fish accompanies your arrival and will again fill the air around here. Until now I was convinced my business was finished and along with it my promise to help my people in Dingle with fresh seafood."

"No need for tears now. We will deliver all you need," Benji cut in. "The captain has agreed to make me part of his crew on future trips, so no more Arne and Stok."

* * *

Aideen swung open the doors of her storage area, and de Portu shouted back to Shicki. "Have the men show our good friend how we fill an empty room with fish. After, we can all head down to Bessie's and celebrate a job well done."

A human chain formed to offload *Espiritu*. Crates and kegs filled with cod were filed in without end. Aideen invited Benji and de Portu in and cleared places to sit.

She was bursting to tell de Portu her own news. "Come and sit here with me out of the way you two. There is so much to discuss. Right after you delivered Father Niels this spring the ambitious young priest mentioned to Vicar Maurice how he could create an important pilgrimage for the local parishioners, leaving for Compostela from right here in Dingle and returning with wine for the parish needs at communion. Maurice sent a message to the Pope in Rome claiming he found a 'higher calling' for Father Niels and requesting he be directed to remain in Dingle to organize such a pilgrimage. He's deciding what varieties he will require the pilgrims to carry back for himself and wants Niels to supervise so that only the best wines are delivered."

"Wait — twice you said Father Niels," Benji interrupted, "the same priest I know? Is he here?"

"Well, yes, maybe the same one, I don't know. Our Father Niels is now in charge of planning this pilgrimage to Spain. Think how busy my cousin Bessie and I will be when pilgrims from all over Ireland show up at her brewhouse to prepare for …."

Benji jumped up. "Tell me where this priest is. I must see him."

Aideen described the tiny crofter's hut tucked into a hill bordering one of the pastures where Niels spent his days. Benji jumped up and dashed through the open doorway.

* * *

De Portu watched Benji scurry away and moved the keg he was sitting on closer to Aideen.

"You know, Aideen, things have changed in my life too, so I have something to discuss with you. I've been a widower for over a year now, and as a boat owner who has passed too many seasons at sea, Vicar Maurice's idea of having me deliver wine from Spain means I'm beginning to plan a new life ashore. What better place than here with you? I can now help you expand your import operation."

The unexpected proposition unsettled Aideen. She skirted the issue by telling de Portu how the abbess in Limerick appealed to the Vicar General in Dublin, requesting that he decree all imported wine to be tax-free. "Not surprisingly, our vicar chose not to tell me anything of it. He let us go on bootlegging and thinking we had deceived him. All to explain we are now free to import at will. Since sales at Bessie's are on the rise, I'll need your new supply of wine. At least there'll be something better to fill this space than swallows' nests."

De Portu brightened. "As I said, with extra deliveries you'll need a partner onsite here. You and I, we —."

"Your wine deliveries will be most welcome, Captain, but we can't forget I need you to fill this storehouse with fish."

De Portu tried to recover. "I'll be pleased to play my part. *Espiritu* will arrive next spring with wine and then continue on to Vinland. When we return, we will have the best of their fish for you. Still, I think you'll need a partner, an experienced operator who will handle the landings from Vinland and manage the fresh kegs of rosé as they arrive."

Aideen's response was quick. "I am hoping to ask the young Greenlander you rescued to do as much, as soon as he returns."

She could see how the mention of Benji as her choice for partner stung de Portu. He hid his dented vanity by suggesting that he go back down to *Espiritu* and supervise the rest of the unloading.

* * *

Niels was shirtless and covered with chaff from bundling the dried hay into sheaves. When he heard a familiar voice shout his name and saw Benji across the field, marching directly for him, he dropped his pitchfork and stood in disbelief.

"Benji! I have been many places and seen many new faces, but never did I think I'd see the face of my one true friend. I was sure you lost your life at sea with those ungodly pirates."

"You worried about what happened to me? I was worried about you. The sinking was Arne's trick to get rid of you and sell the fish. I was sure you would blame me for what they did to you. I did know you survived, but I can't believe you're here in Ireland."

"For that, I must thank you, Benji. The two Vikings in the dory with me admitted you told Arne to make sure I survived. After a herring boat rescued us, I did so much more than survive. Landing in Galicia became a wondrous blessing for me. I lived at the grand cathedral of Santiago de Compostela and then, thanks to Captain de Portu I ended up here, where I discovered my true calling. Few people can understand how deeply my travels changed my outlook."

"It is difficult for me to understand," Benji admitted.

"At first I was motivated by the overwhelming need to feed the congregation in Aarhus. Eventually I came to understand it would be better if the church had fewer Fast Days and less need for fish. For a few similar reasons, I have turned a page in my life." He took a moment to think back and continued.

"During my assignment at the cathedral I greeted the new arrivals, offering them not religious instruction, but more importantly, simple, practical advice. There was one pilgrim ... you can't fathom how many woes this one poor rich lady was burdened with. After hearing her problems and those of all the others, I discovered I am a man who seeks a simple life. You see Benji, instead of living as a parish priest, when de Portu landed me here I knew my destiny was to become a friar. Picture what I have here — no masses to organize, no prickly woolen cassock to weigh me down. I pray that there will be a fine cathedral here someday, but with no list of regular parish responsibilities for me. For now, I have this cottage and my own garden. I walk my own path."

"I see how relaxed you are," Benji agreed. "Should I call you *Brother* Niels now?"

"No, I am proud to be *Father* Niels, but I have become what I do. I tend the crops under the vicar's control, not the flocks of parishioners as Bishop Olsen had me do. I choose to lead a brother's life. Acting as a friar allows me to be a simple sheep in a flock."

"Is your vicar satisfied to have you live a simple life on his estate?" asked Benji.

"I sense he is pleased with me, but many in the village are not pleased with him. He expected Aideen to pay his unfair import tax all summer, even though the Vicar General had ruled it was to be dropped. When word got back to the Vicar General in Dublin, he ordered Vicar Maurice to act more Christian-like. He has tried to, but still he lost the respect of his parishioners."

"You should have his job then. Who better to help his fellow man than a priest who has been through Hell as you have?"

"My recent experience was difficult Benji, but it was no Hell. And anyway, a promotion is the opposite of what I wish for now. Once a year I will advise our parishioners how best to endure their arduous pilgrimage. What I do by preparing the travelers will save the staff at the great cathedral much effort. It will be my contribution to all the hard work they do there."

He buried his face in his hands. "God in Heaven. Listen to me. The Lord is testing me and my pridefulness again. Tell me what happened to you Benjamin. How did you ever survive the brothers' cruelty and yet return to safe harbor here?"

* * *

When Benji recounted how men from his own crew lured him ashore, tied him up and left him marooned, Niels crossed himself and bowed his head in a silent prayer of thanks for Benji's safe return.

"But you are here. How did you survive? Was it the Forest People who assisted you?"

"They call themselves Beothuk, Father, and yes, they welcomed me."

"Beothuk." Niels corrected himself. "Thank you. I should have known."

"Since I was stranded with them, they offered to teach me many things. In truth, if I wanted to eat, I had to get along with the men. And the Beothuk women — I learned much from them too."

"Did you ever see the two lads again — the ones I spoke with on the beach?"

"Do you mean when we first went ashore for water? Yes. They remembered you but only because you acted so silly."

Niels stared up toward Heaven. "I expected them to thank me for the fish we carried in and to open their hearts in gratitude to the peace of our church. Now I realize they already enjoy the peace of the land around them. I'm hoping to find the same."

"You had a terrible experience at the hands of Arne and Stok, but I see the peace you speak of. Its glow paints you from head to toe. I hope to be so fortunate after we repay them for their cruelty."

"Revenge is not for us to seek, Benji."

"Well, my friend, there are ways to balance the scales. I'll explain it better for you when *Espiritu* returns, but now I have to join the crew at Bessie's brewhouse to celebrate with Aideen."

* * *

Aideen stood carefully at one edge, counting the barrels as the crew stacked them high against the walls, covering all the available floor space.

"Captain, all this has finally brought an end to my mischievous dreams. Your crew has filled my space completely. My humble garret is also packed so tight, I can't see my couch-bed, let alone tuck myself in for the night. And something I never imagined — the fish are so plump."

"So, will we now hear the sweet jingle of silver for the crew?" de Portu asked. "Most important for them is to show off fattened purses when they arrive home."

"Yes. We should settle up." Aideen invited de Portu up to her loft and retrieved a strongbox from its place under a loose plank.

"To start, here's a pouch of copper tokens, good for the ale over at Bessie's."

"I was expecting a hard bargain, and I know the crew will thank you for as many copper tokens as you'll give us, but now I'm wondering. Is your next offer to be a stocking full of Viking silver?"

"You should know, Captain, it was a dusty poke full of Viking silver I handed over to those Norse pirates for their fish, and they went away pleased. I guess you'll be accepting nothing other than the finest of Dublin's new pennies."

She dug deeper into her strongbox and set two pouches of shiny new pennies on top of a keg. "I've been assured these are ninety percent pure and newly hammered. Only the finest of silver for you. So you know they are genuine, see the English king's image there and the long cross on the other side. Shake the purse, you'll enjoy the pleasing clink these pennies make."

De Portu bowed graciously. "Perfect. And, if I may, there is one other favor the men have asked of you — a sweetener, if I may. They'll be needing two cart loads of straw for stuffing into their mattress bags. There's been nothing but bickering since we left Notre Dame Bay over the sleeping conditions, and we still have three nights before we arrive home."

Aideen was pleased to explain how the summer had brought extra growth to the hay crop and Niels was in the midst of a second cutting on several of the Vicar Maurice's fields. "I'll ask Poppie to take my cart and find him. He should be in one of the back fields near his crofter's hut these days. He can collect the bedding while you and the men are all at Bessie's."

"After that, we'll be leaving — all of us except Benji. He's a hard-working lad and I should be asking far more than a cart load of straw in exchange for him." Aideen knew it troubled de Portu to admit his advances had been rejected, but she was excited to hear him confirm Benji would stay.

* * *

Benji hiked back to Aideen's and she greeted him with a hug. "I'm pleased you're back. The crew has completed the unloading, and I'll need some help handling it all. You might say I need help from an experienced lad who has an eye for fish and getting it stored away."

With an easy smile, Benji pulled a keg over next to her and sat. "Hmmm. I guess I do know cod as much as anybody on this earth does now. It's why the captain took me on as crew."

It wasn't the response Aideen was hoping for, and she tried again. "Plus, there's my offshore wine sales. I need a lad who can learn foreign tongues, as you did with the Forest People, so you can help when we buy and sell products from foreign boats here."

"The Beothuk's language you mean. Yes, once I learned the basic phrases, I could talk to them and learn many more things, especially from the women."

Aideen tried again. "Well, I mean, if you learned their language, what could you do if you stayed here with us? And my cousin could use help, too. I told her you already know how ale is brewed. You could help her and she has a loft where you could stay."

Benji couldn't torment her any further. He gently tucked a wisp of her hair behind her ear. "Do you remember when I was here with Arne and his crew last year? I told you I would set things right. Perhaps I should call the Captain de Portu over, and we'll tell him his best crewman is staying here to help you."

She held his hand for a long moment and didn't want him to let go. She led him through the crowded storehouse toward her garret. In her mind, she told herself, *People will natter, but this is my place and I'll do what I want.* Then she stopped — saved from temptation when she remembered her private space and cot had been completely crammed full of fish.

"Are we going to tell the captain I won't be crewing with him anymore?"

"Yes, exactly what I was thinking, as we can go over to Bessie's and see if she'll take you on as her new brew master. She needs some help if you're willing."

Chapter 29

San Sebastián, northern Spain

When *Espiritu* reached the breakwater of its home port, the crew hurriedly secured the vessel, bundled up their kit bags and scurried off home. De Portu had a different objective. He headed across the seawall to drop off six trays of salted fish to the owner of *Bar La Navarra*.

"Sancho. It's always a pleasure to see you again. We want you to have these as a gift from the crew to serve along with your fine cider. Leave the fish to soak overnight and serve as fresh the next day," de Portu explained, with a broad smile.

Sancho was fond of fishermen. His establishment depended upon them. And nobody was happier than he was to see de Portu and his crew return, but he couldn't think salted fish would interest too many of my regulars. "I'm willing to try it, Captain, if you say so," he agreed hesitantly.

Later that same evening, fishermen off a Portuguese sardine boat tied at the wharf. Hauling nets in rough seas had left its crew thirsty, hungry, and ready for an evening at Sancho's bar. Sancho offered them a platter of de Portu's surprise gift and waited for their reaction.

The next day when he saw de Portu, Sancho greeted him with the good news. "The Portuguese fishermen picked it clean and demanded more, along with more of my cider. They wanted to know how I could offer such a fine plate of salt fish here in the Basque Country. Captain, I need more information on where you have been. More to the point, I need more of what you can return with next time."

"I can only tell you our journey continued well past Ireland, but who knows if we will be as lucky the next time we go." De Portu knew now that his gift of a basket of salted fish to Sancho would produce a large return.

***

At the back of *Bar La Navarra,* the crew of *Espiritu* stood at their usual tables, waiting for the Cider Moment. As always, de Portu was late to enter. When he did arrive, Sancho cornered him.

"Your men are still being silent about their recent trip, Captain. It's not good for business if they don't look like they are enjoying themselves."

"For the Cider Moment tonight, could you tap a fresh barrel for us?" de Portu whispered to Sancho.

The crew watched Sancho walk over to a bulky-looking cedar tun propped up on a sturdy frame in the back. Each man hurried over to line up for his turn as Sancho streamed the fresh cider through the air and into their waiting jars.

After serving all the men Sancho couldn't wait to tell de Portu more of his experience serving the Portuguese customers. "Captain, I must thank you again for the trays you delivered. The Portuguese crewmen couldn't believe how great it tasted. They wanted more, plus extra cider to go with it. It was the most popular item my humble bar has ever offered. You must get me more so I can create recipes to serve — maybe in an omelet or perhaps fillets with peppers."

"Didn't you guess why we delivered it?" de Portu kidded. "It will be an honor to be your regular supplier. And the news is the same in the town's marketplace too. Our fish is fetching double the price that the butchers ask for their salted whale meat."

"Captain," interrupted Sancho, "The Portuguese fishermen asked where it was from. Your butcher in the market must wonder the same thing. Please, you must give us the details."

De Portu shrugged. "In a way our fish is no different than your cider Sancho — it's best not to share too much information."

"This cider house has been famous for generations and our reputation is hard-earned, but if I don't have any of my own cider, you would go elsewhere. With this cod, no other bar offers such a product. What's the point of keeping this best-tasting dish to yourselves?"

"Some things we Basques keep secret, right? We'll be telling no man and certainly not the Portuguese. They are too efficient at the fish."

"Your men are being too silent, Captain. Why aren't they bragging about where they found such a fortune?"

"I am glad to hear they are being quiet. This will be the most important confidence I have ever asked them to keep."

"I don't understand."

"Tell me, Sancho, do you show the recipe for your fine cider to other bar owners?"

"No. These are things we cider makers are careful to keep within the family."

De Portu smiled. "See, you do understand, my friend."

***

After Sancho walked away, Elkano gulped his cider and confronted the captain with the serious concern all the men had. "We sold most of our load to your buyer in Ireland. Now nobody here has seen how much fish we brought back from Vinland. Nor can we tell them what great fishermen we are, maybe the best ever. How long must we keep the location to ourselves?"

"Forever. If others learn of Vinland, they'll follow. We won't be the best fishermen for long. I understand how keeping this is not your idea of a great life, but my friend, is a full purse not enough?"

Elkano countered. "You don't want us to talk, and I won't be the first, but for me, it would mean something to also be known as the best."

The other men listened to the debate. Elkano asked each of them what they planned to do. Most admitted they wanted to stay on as de Portu's crew for another trip next year.

"Well, for me," Elkano decided, "I'd rather go back for a boatload of whale oil. If I hunt a whale or two here in the winter and several whales off Vinland in the summer, I'll surely be rich, and the world will know I'm the best at whaling."

Only two of the crewmen — the two taken at spear point during the initial confrontation at the Beothuk village, said they wouldn't tempt fate again by going so far and stepping into unexplored lands. Quietly, de Portu pulled the two aside. "You don't want to make the crossing next season, and, for different reasons, I, too, have come to the same decision. For me, I've been offered a chance to transport wines up to Ireland and ferry pilgrims back here for *El Camiño* ... a soft life as a merchantman, eh? If you agree to join me, we'll get a new boat and our coming missions will keep us a little closer to home. Shicki will take command of *Espiritu* for next season."

"Captain, you make the coming season sound better than the last. We accept."

"Keep this offer under your felt caps until tomorrow, men. I need to make sure Shicki is on side."

* * *

When Elkano approached *Bar La Navara* for the Cider Moment the following evening, he had decided to confirm that he had made up his mind to have his own boat built and was going back to Vinland to hunt the huge whales there.

Stepping inside he could see Shicki and the crew already enjoying Sancho's cider, and they each had a platter of his new salt fish omelet. But de Portu was missing and Shicki was buying the cider.

"If it is now you who is buying your cider, Shicki, has de Portu sold out? Are you the new captain of *Espiritu*?"

"Another mug for our friend, Sancho," Shicki shouted, "and yes, de Portu and I have made a deal so I'll be the new captain."

"You're the right choice. I hope it goes well for you."

"We want things to go well with you too, Elkano. If you agree to be my First Mate, we're certain our next season's landings will be greater than the last, more than any fisherman has ever landed. Down your cider and accept to be part of this crew again."

"Word of these omelets will spread. Next time we'll be sure to sell everything right here at home, where all of San Sebastián will see the great fishermen we are. I would only be willing to join you if we bring much more home."

Chatting contently with refilled mugs of cider, Shicki explained they would still need to leave some of their load with Aideen, but he agreed with Elkano to bring most of the next load home to San Sebastián.

With a grin, Elkano announced, "We are already the wealthiest fishermen in town. Next year we'll show the whole world we are the best fishermen ever."

Shicki raised his jar. "Remember, we'll still keep the location of the fishery to ourselves, and there'll be no mention of the whales. We don't want whalers following us. I ask you all to swear this on the memory of your grandmothers."

To a man, they all swore a sacred oath to tell nobody.

Chapter 30

Dingle, Ireland, May of Year Three
Espiritu's arrival was a much-anticipated occasion. On a crisp spring morning after a deary winter, its cargo of new wine would be a welcome boost to Aideen's spirits. The Lenten season was approaching, a time of high demand, and she had only a few kegs left of de Portu's Basque wine delivered on his last trip. But there was no de Portu. Aideen's great surprise was seeing Shicki standing solo at the helm of *Espiritu*. Gone was the oversized vest he sported as First Mate, replaced by a tailored leather doublet. His hair was now close-cropped and completely tucked under a new beret of reddish-brown felt. He stepped down onto her wharf as soon as the vessel was secured.

"Soon after we arrived home last fall, Captain de Portu announced he planned to turn his long-standing investment in *Espiritu* into gold — golden doubloons, in fact. He received fair payment, and I'm the captain now. Elkano, there behind me, is my First Mate. No worry for you though," he added. "We will still be your supplier of all the wine you need from the Basquelands, plus a measure of what we find in Vinland."

Aideen was concerned by the distinct phrase he chose.

"A measure, Captain? I enjoyed kind relations with Captain de Portu. I hope you and I will have the same arrangement."

"First, you must still call me Shicki. There is no need for 'Captain'. I must tell you, though, this year there will be a change in the arrangement. After we tied up at home last summer, word of our special cargo spread quickly. Demand for salted fish ballooned. The owners of the cider bars wanted all we could deliver because it makes their customers thirsty for more drink. The peddlers in town also wanted more than we could provide."

"Is all this by way of saying you're keeping a greater share on board and for yourself this season?"

Elkano was standing behind the new captain listening. He jumped into the conversation with a blunt, "Yes."

Aideen frowned. "So tell me, Shicki, if the demand in San Sebastián is so strong, why not encourage more of your compatriots to join you across the sea?"

Again, Elkano's response was blunt. "We are rich fishermen now, but we still want to prove we are the best fishermen in the world. To do so we need to return home with more than anybody else has ever seen on one fishing vessel. Unfortunately, to do that, we will be leaving less here with you."

***

As soon as payment for the wine was handed over, Shicki had the vessel's lateen rig dragged out of the locker and rigged for the crossing. Aideen was pleased with the wine deliveries, but couldn't ignore her deep disappointment as the vessel departed. She sent Poppie to fetch Benji, who arrived in time to see *Espiritu* slowly disappear.

Benji sensed Aideen's unhappiness. "We'll see a fine profit from this wine and strong sales for whatever amount of fish they bring back for you, but it's plain Shicki and Elkano have left you troubled. What can I do?"

Aideen took his hand and led him over to an empty section of her warehouse where they could chat quietly. "Can you tell me why, with such a strong demand, Shicki and Elkano would want to keep the fishing grounds of Vinland to themselves?"

Benji had a simple answer. "Fishermen don't share their grounds with any other crews if they don't have to."

"Why not? You said there is no end to the fish over there, and we know there is no end to the demand here."

Benji swung his free arm back to the farmlands across the harbor. "Aideen, if the crops over there are thriving, the farmer will expect to harvest all the beans he can see because his crop doesn't move. On the other hand, fish have tails and they swim. No fisherman can see deep below the surface, and he can't count on anything until he has it stowed on board. It's sure Shicki will locate the cod in Notre Dame Bay, but no true fisherman ever trusts what he can't see."

"Still, what could it hurt if a dozen vessels from the Basquelands arrived in Vinland? You have said there would be more than enough for them all. They could all show off the largest landings in San Sebastián."

Benji gripped her hand gently. "You have to understand. The Basques are the world's best whalers, and I suspect they will be the best fishermen anywhere for years to come. For them, the larger the treasure, the more it must be protected."

Aideen stood up so she could stamp her foot. It made little noise, but her reaction reminded her of how she felt the day she was dismissed from service. "I once promised to have good fish available for the people of Dingle, and if it means sharing the location of Vinland with anybody who has the courage to head over there and bring it back to me, then that I intend to do."

*　*　*

Since last summer, the quiet village of Dingle took on a new vitality and gained a reputation as the single best place to eat and drink in County Kerry. Though it was a long way off the main roads, Aideen's growing offshore trade and Bessie's busy brewhouse gave people a reason to spread the county's reputation for hospitality.

The situation irked Maurice as he sat at dinner with his retinue of cronies. "I have said before, women should not be running businesses such as those, but it's women who run them both, and successfully."

"And it gets worse, Vicar," added one of his guests, quick to alert him of *Espiritu's* recent visit to Dingle. "I'm told the vessel dropped off many more kegs of wine."

Maurice's face reddened. "The church expects all good souls to fast during the Lenten season. We must pray and repent. With plenty of their codfish and now more wine, the Lenten period for my parishioners has become a time to enjoy themselves. Nobody thinks of doing without."

Maurice received some satisfaction when word came from London of a new tribute payment for the king. It was to be paid by all commercial ventures, including Dingle's two thriving establishments, and local vicars were to collect it. The chance to collect a new levy from both the village fishwife and the alewife made Maurice smile. He called for his horse to be readied.

The ride to Aideen's storehouse took him through the flowering fields of Beenbawn, but with his chronic problem with piles, Maurice was in agony. When he arrived, he tossed the reins to Poppie, slid from his saddle, and spat the road dust from his mouth. Poppie whooshed at the pungent fragrance clinging to the air around him, then hid his amusement at the stories of the vicar bathing regularly in rosewater. He ran to fetch Aideen. "The vicar is here, m'lady, and he smells worse than the butt end of his horse."

* * *

Peering through the open warehouse doors, he could see kegs of wine lined up against the walls, each waiting to be inventoried.

"I heard your ship has come in," Maurice declared when Aideen arrived. "Thanks to your efforts in high places, I no longer collect duties on valuable imports such as you receive, but *you*, on the other hand, will earn a sizable profit from all this."

Aideen ignored the taunt but suspected there was a hidden reason for the vicar to be standing at the entrance to her warehouse. Her response was civil. "M'lord, if you're thinking there is enough wine here to do much more than pay for its delivery, let me set you right. There will be little profit made this season. You should be satisfied I am able to pay your outrageous rent when my ship returns from Vinland."

"How rewarding for me." Maurice's sarcasm belied his smile and pleasant retort. "With all I am seeing here, your venture seems to be thriving well enough. Strong returns from a thriving enterprise mean you can afford my rent and, dare I say, an extra contribution. You see, there has been a royal decree. The king's court is imposing a tribute levy upon all successful entrepreneurs, including you. The funds collected will pay for upgrades to King's Castle in Limerick. After all, those formidable castle walls are for your protection. I believe you have been there. Surely you have appreciated them."

Turning on his heel, he grabbed the reins from Poppie and climbed gingerly back up into the saddle. Over his shoulder, he tossed a parting missive. "The king's levy will be due, along with my rent, on the day that ship of yours returns. God save the king."

Poppie stood beside Aideen as Maurice disappeared down the road. Seeing the slump of her shoulders and the frown on her face, he turned to her and began a new rhyme.

The vicar of Dingle
rode out this day,
intending to —

Aideen waved her arm. "Sorry, Poppie. Not now. I'm too upset for your rhymes today."

* * *

Aideen woke at dawn but in a mood more sullen than the one she took to bed the evening before. She walked to Bessie's for a bread bowl full of her cousin's special pottage. Bessie placed her bowl on the table and dropped down beside her, crossed her arms and exhaled heavily.

"What is all this for?" asked Aideen. "Are you upset by the new tribute payment, too ... the one our vicar has imposed on local businesses?"

"Yes, I am upset, but it's not only the tax. Your Benji was supposed to be a great help for me and my husband, but instead of helping us with how we do things, he wants to change my recipe for ale. With Lent upon us he's telling my best customers all about his dark brew. 'It would be a better choice than the ale on offer here,' he says. Better than your Basque wines, too, he adds."

Bessie was more than proud of her recipe, always explaining it was the best choice of any in County Kerry because it served as a complete meal. "After you empty a jar of my ale, you chew the bits of barley still swirling at the bottom. But Benji has taken to smoking his barley first. He makes the drink stronger and darker, but it leaves no pulp in the jar. Where's all the food value then? Aideen, my cousin, this is a problem for me, but one you have to help solve."

Aideen never considered ales could taste different, one from the other, but did remember how Arne and Stok once told her Benji's rich brown beverage was stronger and had more flavor than Bessie's weak brew. She tried to calm her cousin.

"More choices might bring you more customers. Think of Benji's recipe as the key to doubling your sales. Explain how your traditional ale, with all its chewy bits, is more than a drink. Promote it as a meal for midday. Tell them Benji's dark brew is the stout refreshment they need for later on, after their work is done. And be sure to mention it will help them with their sleep."

"I don't think too many of my lazy customers ever have trouble sleeping, but you make me see how it might double my sales. If I have a new and different ale, it will need a new name. Customers will come to know it as the special 'Stout Ale' from down at Bessie's brewhouse. Travelers from afar will follow the dusty road all the way to stop in here for Bessie's Balls o' Fish and stay for a few jars of the new Bessie's Stout."

As Lent went on, Bessie was pleased to remind her customers, "Lent is no longer a penance of oily herring and thinned-down ale. You needn't suffer through the fast. We have the tasty meals you want and they've all received 'Papal Approval'." She also boasted that the ale she was serving was too popular for her and her husband to enjoy themselves. "He and I, we won't drink the pennies out of our own purse," she explained.

* * *

Two months to the day after Shicki and Elkano left, Aideen sighted *Espiritu* approaching the harbor. The vessel was riding low, and with no hint of wind, she knew it would be a while before they could tie up. She asked Poppie to fetch Benji from Bessie's to help her make extra space in her warehouse for the anticipated load of fish.

With the vessel finally tied at her wharf, Aideen unfolded her wimple, layered it over her shoulders and shouted up to Shicki. "I'm pleased to see you and *Espiritu* back on our side of the sea, and I'm particularly pleased by the pleasant aroma accompanying you. Does it mean your voyage was the success you planned?"

"Success is not enough of a word to describe this load. There's not a rat hole aboard without a fish stuffed into it."

Aideen was hoping his comment meant he had what she needed. To encourage him, she offered an extra sweetener. "I set some extra copper tokens for you and the crew. After you've finished unloading, Benji will lead you back to Bessie's for a few bottomless flagons of his new stout amber ale."

"She's right Shicki," Benji shouted. "I have a dark, creamy brew on right now and I have to get back to draw it off, but —."

Shicki interrupted him. "You make a tempting offer, Benji. We'd all enjoy a jar or two of your ale, but as soon as Aideen's share of this load is on her dock, we're leaving."

"I ah ... since you're right here, I was hoping you might at least take a minute to talk. We both want to keep more because our customers are clamoring for it. With all you will be landing at home, don't you feel you could let an extra boat or two share this treasure next season? It's not as if this is the biggest secret in the world, is it?"

"Yes. Yes, it is. We'll keep those others guessing, and you will have to be happy with what we can supply, while we enjoy landing the most fish ever seen in San Sebastián and be known as the best fishermen in the world."

* * *

Espiritu turned and headed offshore, still fully loaded. Aideen and Poppie agreed it was sitting as low in the water as when it arrived. The two ambled back into the warehouse noting how the space was nowhere near full enough to carry them through the season.

"I'm sure those cronies of Vicar Maurice squealed louder than magpies when they saw our supply ship here today," Aideen grumbled. "He's likely thinking my storeroom is packed full again. Sure, he'll be back expecting his rent — plus the king's extra levy due now for the grand castle in Limerick."

Not a moment later, Poppie announced, "I suspect that cloud of dust I see down the roadway there is him coming this way on horseback, mistress."

Maurice rode right up to the storehouse and dismounted. Not bothering to hide his discomfort from sitting in his saddle, he walked directly inside. Any idea of collecting both his rent and the new tax evaporated when it became clear to him there was nowhere near enough product in the room to satisfy the local demand. Undeterred, he spotted Aideen and marched over to her.

"I'm here to collect my rent and to remind you of the new levy the king has demanded of all business owners. With your season's supply now arrived, I would think this is the right time to make my collections."

"Look around you. There'll be no profit made here this season." Aideen stamped her foot. Her effort vibrated the floorboards across the empty storeroom. The rafters shook, releasing dust, feathers, and dried guano from the swallow nests.

"The vessel leaving our harbor dropped off less than a quarter of the fish my people need. And it won't be back this year. Nor will any others. I'll be lucky to scrape up your outrageous increase in last year's rent."

"Aideen, you might claim you are having a poor season right now, but when I observe this bird-infested knock-up you have built, I can only conclude your costs of doing business are minimal. You've surely been squirreling away the past wealth you earned. I'll go so far as to guess there's quite a reserve tucked under your straw mat."

Aideen was flummoxed. "You see how I live. This is no grand manor we're standing in, but at least it's my own straw mat I sleep on, in my own modest home. The only good thing is that if you hadn't dismissed me two years ago, I would still be your housekeeper, with nothing at all to call my own."

"I'll not be unreasonable. You may pay me your rent now, but as a personal favor, I'll wait for the king's levy to be paid in full on All Hallows' Eve."

Aideen fetched five shilling coins from the strong box up in her loft and handed them to him without comment. Tossing off a self-satisfied glance, he turned on his heels and mounted his palfrey.

Watching the vicar trot down the road, she called Poppie over. "I think I'm ready for one of your rhymes to cheer me up now."

"Well, Aideen. Here's the one I had ready for you a while ago."

The Vicar of Dingle rode out this way, intending to fill his purse.'

'I'm here for your rent,' he said, 'but these piles are the real curse.'

'I soaked them in rosewater, but still I twitch.'

'It's like there are thorns on the roses and their pricks make it worse.'

Chapter 31

Dingle, Ireland

Niels was in the vicar's high meadow, stripped to his britches and pitching a new cut of sweet-smelling hay into tall mows. The constant onshore breezes had been perfect for drying the crop, but this work was making him sweat and the chaff stuck to his back and hairless chest. He lifted his head, wiped his brow with a bare forearm, and stood facing the horizon. In the distance, he spotted a tall ship, one different from any others. *It looks to be shorter than a whaler, and its rigging rises too high. It's more of a coaster*, he said to himself, but then he remembered the arrangement Captain de Portu made with Vicar Maurice the year before.

"Ah, welcome back, Captain," he shouted into the wind, then poked his pitchfork into a haystack and pulled on his tunic.

Niels ran along the shore parallel with the vessel's progress, headed for the shoreline and Aideen's storehouse. When he arrived, he saw she had also sighted the unfamiliar vessel approaching and was preparing for its arrival.

"Do you know who this might be entering our harbor, Aideen?"

"It's not *Espiritu*," she declared with certainty. "At first, I hoped it was another caravel following Shicki to Vinland, but it's too late in the season for a crossing now. What could this foreigner want from us? I can't be sure yet who's standing at the helm, but he handles his vessel as if he is familiar with the approaches to our harbor."

* * *

The vessel neared the wharf, and the captain called over, trying to conceal his voice.

"Hello, *andrea*. And greetings to you, *Pater*. I am the proud owner and master of this new vessel. May we tie her up here and come ashore?"

Aideen could hardly control her joy. Before she could remember any of the Basque words of welcome she once tried to learn, de Portu rushed down the gangway and lifted her in his arms.

When her feet touched back down, Aideen blushed. "You can't play any more of your tricks on me, Captain. Nonetheless, 'tis a pleasure to see you and this new vessel. Many things must have changed for you this season. I was told you gave up your life on the sea, yet here you are, and with a new vessel. And it doesn't smell of sheep dung."

De Portu called over for Niels and dragged him into the three-way embrace.

"Father, did you not let Aideen in on our plans?"

Aideen interrupted, "What? Have you two been planning a surprise behind my back? Since you don't need to sneak your *communion* wine in anymore, this must be another scheme to tease me?"

"The captain and I spoke with the vicar last year," Niels admitted. "We worked out an arrangement to deliver the Spanish wine he fancies. As for me, I'm to lead the first church-sanctioned pilgrimage to Compostela from this harbor. And it won't be the last. This will be a real blessing, allowing me to help the unprepared discover the true peace of the sanctuary there."

"I'm glad for you," Aideen said. "Surely you deserve this."

De Portu chimed in. "I had this fine new vessel built with those passengers in mind, the ones the vicar will select for his pilgrimage. More importantly for him, she's perfect for delivering a cargo of wine. I ordered an extra lateen sail too and explained it would be needed for the swift transport of both the pilgrims and his wine. The pompous oaf agreed to pay for it."

De Portu smiled, caught his breath and held up his hand, as if to signal he had a special announcement.

"And while it was being built, I researched your Irish expressions. I wanted a name for this vessel to honor you. She has been christened *A Chara*. I have been told the meaning is 'my fine friend'. I hope this is alright."

* * *

Aideen was reminded of how she spurned de Portu's advances in the past and changed the subject. "Captain, this will be another busy season for us after all. Since my rafters already bulge with the wine Shicki delivered, why don't you and Niels join me to tap a keg and celebrate? I'll send Poppie to fetch Benji."

She rinsed out pewter mugs for each of them, and they found kegs to sit on.

"I once promised myself this would happen," she said to Niels, "and it is indeed a blessing for me, but there is one other important thing."

"Should I be worried you are going to ask me to act against my vicar's wishes, Aideen?" Niels asked, more than half-seriously.

"No, the vicar is not our biggest problem here. Pilgrims and wine are all good for business, but we need more fishermen to cross over to Vinland. Our people want far more fish than Shicki will share with me."

"I understand you are not thrilled with the load Shicki left for you but perhaps you should count your blessing to have what you have."

"I don't mean for me here and the people in Dingle. Think of all the hungry parishioners in all the Christian lands on this side of the sea. You must speak to a master fisherman in Spain or maybe your contact, the Queen of Portugal. Tell them all of this chance to develop an outstanding food fishery."

"Perhaps the task is one you should more properly address, Aideen." Niels knitted his fingers in front of him. "All your friends have visited far-off places — buying wines in Spain, or living with the Beothuks in a new land." Niels took her hand. "Do something for yourself, Aideen. The best thing for you is to join us and deliver the message to fishermen outside Dingle."

"I'm content here. I don't yearn to be anywhere but on this shore, particularly after the overnight trips to Limerick I took with Paddy in his boat of animal skins."

De Portu joined in. "I agree with the priest. The sea air can offer many favors. It's been two years since you promised to build on this site. You have succeeded beyond measure, but you've not moved one step away since. You must join us on the pilgrimage to the cathedral."

"To be honest, I've little interest in religion since Vicar Maurice dismissed me. Nor do I yearn to go traipsing over the Galician foothills for days and nights. How would following The Way of Saint James make me a better person? I'm sorry, Father, but I speak the truth of it."

"I understand what you are saying, but the trip can be more important for other matters, not just the church. In Spain you could speak to the fishermen whose boats can bring back the fish you need."

"And my affairs here? How can I leave Poppie by himself with all this?"

Niels waved a hand at all the kegs of wine around them and smiled. "My dear sister, one lesson I learned in Spain was that these kegs Shicki left for you are too young for tapping. If you leave them here to mature, in a month or two, their contents will improve. Besides, you agreed at the time that fish, not wine, was what you needed. If you come with us, you can spread the word to the foreign fleets yourself. Talk to those boats who can deliver a good amount of fish to you over the coming seasons."

Aideen peered out through the louvered shutters to the harbor entrance. "I'm getting advice aplenty now. In all the time it took me to build this business, I was content to do it on my own. It worked for me, but I have always wondered what my shorefront looks like from the visitor's point of view. Is it welcoming for vessels approaching from the sea? If I do venture out to the harbor's headlands with you, I'll get a sailor's view of what my hard work has accomplished."

* * *

Benji was drying his malted barley behind Bessie's when he received word of de Portu's arrival. He had developed a plan in his mind to deal with Arne and Stok, and this was his chance to solicit de Portu's assistance. He doused the brazier, told Bessie he had to see Aideen, and left running. On the way, he rehearsed his request to de Portu. When he arrived, he was more than impressed with the new vessel lying at Aideen's dock.

"Tell me Captain, do you remember last year when I told you how wicked Arne and Stok had become? 'Worse than foxes in a hennery', I think you agreed. The time has come for us chickens to repay the favor."

Niels frowned at Benji. "Have you been plotting revenge my friend? I too, wish we could make Arne pay for his sins, but we have no idea where he is these days. And he is far too dangerous a man to confront. Not to mention, revenge is not for us to take?"

"Don't think of this as revenge, Father. This is how we will teach a pair of evil spirits that they harvest what they sow. Aideen has reminded me that All Hallows' Day is right around the corner. As a priest, you know we must confront the demons on such a night and extract a payment from them for their sins."

"I shouldn't hear any more," Niels stated flatly. Walking away, he tossed a blessing over his shoulder. "May you have the protection of the righteous."

Aideen gave Benji a skeptical glance. "Benji, I enjoy having you near too much to chance losing you again to those pirates.

Don't you dare give them a second chance to take you from me."

Benji took her arm in his and turned to de Portu. "Captain, I've worked out a quick way to make the Gunnarsson brothers pay for the suffering they caused us — Aideen, Niels, and me. I have been to where they hide their silver. It's an isolated island in Stavanger Bay. A fast ship is the final piece I need to complete the plan. Arne and Stok won't be able to tell we've been there and relieved them of their stash until it's too late."

"And there's no chance you will run into Arne and Stok there?" Aideen asked Benji.

"Even if we do, it won't matter if *A Chara* is faster than their Viking longboat."

De Portu responded with unrepressed pride. "*A Chara* is faster than any longboat and as for me, I'm eager to prove it. But remember I told Vicar Maurice we would leave here with his pilgrims in eight days. Stavanger Bay is not far up the coast of Norway, but to be back in time, we'll have to depart right away."

Chapter 32

Stavanger Bay, Norway

Several times during their two days following the coast to Norway, de Portu reminded his crew of the great risk their mission involved. "All Vikings, including those past their prime, are dangerous." Regardless, each man wanted the thrill of sneaking into an actual Viking hideaway, taking all the treasure they could carry, and living to tell the tale.

Before noon on the second day, Benji directed the helmsman to enter Stavanger Bay and turn south around the island at the mouth of the inlet. The bay was flat, calm and speckled with numerous islands. Only a few local boats dotted the horizon, but still, de Portu insisted on a constant lookout for any sign of a longboat. Benji agreed. "We can't be sure if the Gunnarssons' crew is somewhere in the area Captain. I would double the watch."

The wide bay narrowed and Benji pointed at a grass-covered rise on the left. "This is where we need to turn sharp right and find the smaller island all by itself in the open," he told the helmsman. The crewman stared back at Benji with a bemused smile. "All the islands I see out there look the same young man and yet you expect me to pick the right one?"

"Arne chose his particular island because it's all by itself in the middle of the bay. The location is good for us because if they are already there, we'll see their boat on the beach and turn tail in a hurry. The trouble is that if they are not on the island, nobody can sneak up without being seen. His spies might also see us clearly."

Benji stood at the bow, scanning the bay. "I see it," he whispered, pointing to a single tree-capped islet. When they approached, Benji looked for the shorefront where the longboat usually grounded out to put the men ashore. "If their longboat was here, its keel would leave a mark in the gravel." He swallowed hard when he saw the faint skid marks, but was unable to tell how recently the longboat had been there. *Hopefully that mark was made earlier today and they won't need to be back for a while*, he said to himself. He said nothing to the crew.

* * *

A slight breeze danced across the bay and de Portu noted it was nowhere near enough wind for his heavy vessel to respond quickly if they needed to 'turn tail in a hurry'. Continually marking the depth below the keel, he edged *A Chara* as close as he could to the island and dropped anchor. The dory was launched with three of his most capable crewmen. Two of them grabbed the oars. Benji jumped down to the dory, and de Portu held out several rawhide sacks.

"You'll need these for the treasure," de Portu offered.

"There's the funny thing. Arne always separates his cache into individual sacks. It'll make it easier for us to lug the valuable ones back aboard. This should be quick."

The dory nudged up against the gravel and Benji jumped out. The other three hurriedly secured the oars and hauled the small boat up the beach. Benji could sense the men's anxiety as they stepped ashore. They cast careful glances all around before Benji led them up a narrow path into the trees. It ended in thick underbrush where Benji pointed to a steep rise and told them to watch for a rock cave. As soon as the men spied it, they broke into a dash for the opening.

"No, no," Benji warned. "Not there. It's a decoy. Arne planted the entrance with stinging nettles so don't go too close." He waved his hand and pointed higher. "The Viking plunder is stashed in a crevasse above the cave and camouflaged behind two tall trees. I'll go up, you three stay here and keep watch."

He climbed up above the cave, found the crevasse and lowered himself into the narrow cleft in the rock. There was room for one person. He felt around with his feet until a soft 'tink' told him he had kicked solid silver. Tension squeezed his chest as his arm explored the darkness. Minutes later, he appeared only long enough to toss two sacks full of delicate altar pieces down to the three men.

Again, he entered the tight space. Feeling around blindly, he grabbed smaller bags, listening for the clink as they moved. They would be full of small objects, such as goblets, chalices, rings, and cutlery. He located a sack of bulky candelabras but discarded it along with numerous heavy pouches of Norse coins. He climbed back down and told the others, "We've got the good stuff. There's only a few sacks of Norse coins left. It's not worth lugging back to the boat because it has a poor silver content."

The four men lugged the bags back to the beach. Benji could see the wind had come up and the bay was growing choppy, so he spread the sacks evenly over the floor of the dory. Struggling to push their heavy craft out into the rolling breakers, each man whooped and cheered at their success. They jumped in and rowed as swiftly as they could to *A Chara*. Benji too, began to cheer until he spotted a broad, striped sail on the horizon. Arne's longboat was making straight for the island ... and for them.

All sixteen Viking oars pulled the longboat ahead with swift, powerful strokes. Memories of the wicked things Arne had done and could do flooded into Benji's mind. He tried to get de Portu's attention by waving while he shouted for the others to pick up their stroke.

De Portu had already spotted the longboat and bellowed to his crew, "Raise the lateen rigging. It will be our one chance to pick up speed. Drape the cargo netting over the stern to help Benji's crew climb aboard." The crew jumped to the ready as much to keep from panicking as to follow de Portu's order. "Don't weigh the anchor until the dory is secure," de Portu added.

* * *

With the four men and their sacks of loot aboard, de Portu shouted for the crew to haul the anchor and cut the dory loose. He called for the rigging to be set for maximum speed, took the helm and steered his vessel straight for the approaching longboat. The caravel was slow to gather momentum, but the rowers in the longboat were stroking at their top speed. A wide white wake trailed behind them giving proof of their speed over the water. De Portu was amazed the longboat could move so quickly but pleased to see it because he knew their combined speeds would be critical when they met. "I don't plan to collide with them," he shouted. "We are going to sideswipe them, scrape along the gunwales of the longboat and wrench the oars loose from the men's grasp."

When the two vessels drew close enough for de Portu to see the frenzy in Arne's dilated eyes, he swung the tiller hard. "Hold on."

A Chara's prow lurched to one side as her high-sided stern slammed broadside into the longboat. The sound of the first oar shattering was louder than lightning splitting a pine tree. De Portu expected that, but the staccato cracking of seven more oars on the same side of the longboat, plus the steerboard, startled him as echoes rang across the bay.

The longboat was still gliding forward but with no control over steering or stopping. What nobody expected was how the collision scuffed the patch off the longboat's hull, opening the old gaping hole.

Seeing the crippled longboat brought de Portu some relief until both Arne and Olaf leaped from their deck and grabbed the rope netting still hanging over *A Chara's* stern rail. The two frantic warriors scrambled up the netting toward the deck. De Portu saw each brute carried an ax and dagger in his belt. As owner of a vessel built for ferrying pilgrims and wine, he had no weapons aboard. "Watch out. They're armed," was all he could say, but delivered the warning with genuine fear.

Benji sprang to the cutlery bags and dug out the one item he could think would serve as a weapon — a hefty serving fork. He got back to the rail as Arne's right hand curled over the top. With fear propelling his swing, Benji plunged the tarnished but still sharp meat fork through the Viking's fat fist, pinning it to the oak rail. Arne went crimson with rage. His nostrils flared, revealing long black hairs, and his eyeballs bulged. Olaf was right below him, yelling to keep climbing.

Arne swung his left hand over to grab the fork, but Benji pulled it out before the man could make contact. In the half second with only his injured hand holding the rail, Arne's grip failed, and he fell backward. On the descent, he bounced off Olaf and the two splashed into the chilly water below. Benji dropped the serving fork and hauled the cargo netting up onto the deck.

Arne gasped, flailed his arms, and struggled to raise his head. Splinters from the pine oars, many the length of a man's arm, drifted around him, but there was nothing he could grab onto. His woolen blouse was now sodden and weighing him down. The longboat was drifting beyond reach, listing due to the reopened gash in its hull. Thrashing violently beside him, Olaf was barely able to scream, "I can't swim. I can't swim." On the longboat's deck, Stok lunged out with the tiller and waved it, but it was too short. The chilled water had weakened both Olaf and Arne, and neither could swim for it.

Benji kept a cautious eye on the men floundering below as the distance between the vessels widened. "They won't last long in the water, none of them can swim. And it won't matter if they do make it back to the boat. She's done. It's well down to one side, and her nose and figurehead are already below the waves. The entire crew will soon be with Olaf and Arne and a long way from shore."

With *A Chara* picking up speed and heading for open water, Benji relaxed enough to glance over what they had escaped with. He calculated the silverware, once melted down, would earn more than required to cover the new business charges the English king levied on both Bessie and Aideen. Then he remembered the serving fork lying on the deck.

"I want to keep the fork," Benji said, "but Niels should have first pick of any sacred vessels and ornate cutlery for the sanctuary of a new chapel, if one is ever built for him." Nobody heard him over the cheers of *A Chara's* crew as the longboat disappeared beneath the waves.

De Portu waved an arm and called over to him. "Rejoice with us, my Greenlander friend. Arne won't be your problem any longer."

"No, I'm not worried, Captain. And you know we recovered more than enough to reward your crew for their courage today, although every man jack among them will value telling the story of this day far more than any silver we could offer them."

Chapter 33

Leaving Dingle Harbor

Two days after *A Chara* and the men returned from the most daring escapade anyone could have imagined, Aideen climbed the gangway to join the twenty-four apprehensive pilgrims on the main deck. She took Benji's arm and watched as de Portu and Niels discussed the coming voyage.

The idea of being part of such a novel adventure thrilled the crofters, but none looked forward to long days and nights on the open ocean. Several reacted nervously at the slight movement of the deck while it was still tied at the wharf. De Portu took Aideen's hand and stepped into the crowd. "You know, the old Basque sea captains say having a woman onboard a working vessel will anger the sea gods. In truth, since our Aideen here was the inspiration for my new vessel and she is Irish, the sea gods say fair winds and calm seas will accompany us all."

"There are no 'sea gods'," Niels said to the pilgrims. "And I do have more practical advice to assist you with the conditions at sea. Spread your arms. Feel the salt air flow deep into your chest. Your voyage will be a truly memorable experience and you'll want to spend most of your days up here on deck in this fresh air. It's the best place to alleviate seasickness."

Moving out into the current, most of the pilgrims ignored Niels and filed down the ladder to lay claim to one of the tiny bunks or to play dice. Aideen stood at the rail looking back at the sea cliff where her father used to carry her on his shoulders. "I wonder if he ever stood on the shore, picturing his daughter going to sea on a vessel named for her."

***

Benji and Niels also remained on deck, where de Portu outlined the route they would take to Spain. "Vicar Maurice suggested San Sebastián as the beginning of the pilgrimage. It is a traditional starting point for pilgrimages, not to mention it is a charming Basque harbor town. I'll set a course for there."

Niels raised his hand and asked, "May I ask a tremendous kindness for these good Irish souls with us? The vicar did say The Way of Saint James is a mandatory faith journey, and yes, he did instruct them to walk the full distance over the northern mountain range. However, I wonder if you could change direction slightly and land in a Galician harbor to be much closer to the grand cathedral? It will avoid hiking many leagues over a long and arduous route."

De Portu protested, "And bypass my home port of San Sebastián?"

Niels regretted that he might have wounded de Portu's pride and tried to apologize. "It is unquestionably most pleasant and I did enjoy my stay there, but to be truthful, my mission is to help pilgrims find their own way, not have them suffer trying to follow Vicar Maurice's way."

"If such is your wish, my vessel is at your service. We could head for the old port of A Coruña. I think you are familiar with the harbor."

The captain conferred with his pilot. The ship bucked when their heading was altered, and several passengers scampered up on deck and hung their heads over the rail.

The captain went back to Niels with a smile. "Your passengers will not appreciate it, as they will be spending a little more time at sea, but you are right to make this course correction with them in mind. The trade-off is a much shorter walk. And anyway, several of the wines your vicar asked me to retrieve are produced on the clay slopes above the village of A Coruña. This will work out, I tell you. If we follow this route each year, I mean landing at the trail to the cathedral via A Coruña, we should christen it the Irish Way."

"I have one other request, Captain," Niels said, "but it can wait a day or two."

* * *

Niels had forgotten how close the Spanish mainland was to Ireland. In less than a day and a half, de Portu announced the sighting of the famous Roman lighthouse and their imminent arrival at A Coruña.

"You should bring the rest of your pilgrims up on deck, Father. I know you've seen this ancient light before, but you should have the others see it. It's called the Tower of Hercules. The Romans built it right out on the tip of the continent to remind their citizens that Rome ruled all the land out to the very edge of the world. Local fishermen still fear going too far beyond the beacon. Let them stay on shore, I say. For me, the light at the top points to a fortune on the other side of this sea for those who dare."

Niels went below to fetch Aideen and the troop of pilgrims. They all climbed back on deck, as thankful to leave their reeking bunks as to see the calm waters of A Coruña Harbor. Niels rehearsed his second request.

"Captain, there is one other thing we should discuss. The Portuguese queen told me her navigators have no fear of leaving the sight of land behind, if there is somewhere to go that is. When they hear of Vinland, they will be eager to leave their home shores far behind. So, I have a favor to ask of you, a favor for Aideen. While I am occupied with the pilgrimage, I ask that you continue around to Averio on the coast of Portugal. It is home to a fleet of capable fishermen and it's not far from Coimbra where Queen

Matilde promised to deposit my manuscript, although I suspect Her Highness did not do as she promised. Still, if you will take Aideen and Benji to Averio, they could deliver a copy of my chart and tell the fishermen there of all the fish in Notre Dame Bay."

De Portu looked hurt. "But it is my Basque countrymen, including Shicki and Elkano, who are the accepted masters of the oceans. We speak the same language as the cod fish you seek. What could the Portuguese do for Aideen that they cannot?"

"That answer is simple. Shicki and Elkano are Basque. They will not support Aideen with the fish she needs."

De Portu frowned as he calculated the added distance around the coast to Portugal; then he remembered that Averio had a reputation for vast salt lagoons.

"On the other hand, once the captains from home do become active in the distant fishery, much salt will be needed. I will make a fortune distributing Averio salt to the Basque fleet along the San Sebastián coast." His smile grew as he exclaimed, "Selling salt will be a rewarding new career for me, plus delivering wine and ferrying pilgrims."

Once secure in A Coruña, Niels offered an inspirational blessing for the trek and led the group into the low mountains, following the path he had followed a year before. De Portu quickly readied *A Chara* for the next leg of the journey.

* * *

Clouds had descended on the Portuguese coastline and seemed determined to remain, obscuring the shoreline from the sky. Aideen squinted into the fog as de Portu located the outer channel of Averio, Portugal's busiest port, and carefully eased into the inner lagoon. The captain asked the crew and each passenger on board to watch for mud banks and to warn the pilot of exposed salt flats.

As they edged closer to the fishermen's quay, she could see colorless homes clustered around the port and back several leagues up into the wooded slope behind. "It's as if we're looking through a linen curtain and it's stripping away all the color," she said. But as they edged closer, she noticed all the red roofs and the hand-painted tiles on the buildings.

"Their houses are so much brighter than ours. And do you smell the village? It has the same pleasant perfume of salt fish I know from home. I'm over my seasickness, and the perfume in the air has revived my appetite. It also makes me homesick."

A handful of Portuguese fishermen, impressed with the visitor's navigation skills, gathered at the wharf and secured *A Chara*'s mooring lines. As soon as de Portu planted a boot on their wharf they questioned why a Basque vessel, not rigged for fishing, might so skillfully be entering their port.

De Portu suggested Aideen and Benji set off into town. "I'll see to the boat here for now. It's still early so I'll chat with this crowd for a while and wait for the rest of their fleet to return. Why don't you two locate the nearest inn? I'll be along and I'll bring a basket of our Vinland fish to show them."

* * *

Following the fresh paving stones of the main street, it wasn't long before Aideen spied a two-story cottage with a hand-carved sign over the entrance indicating food and lodging. Linen curtains hung in the window, fringed with intricate tatting and tied back to allow a welcome glow to shine through to the street.

Crossing the threshold, they expected to be greeted by the familiar tang of stale ale, but the zesty fragrance of young wine filled the air. Most of the tables in the room remained empty except for a few locals playing dominoes and enjoying glasses of a cloudy, green-tinted wine. Aideen and Benji chose a table near the door.

A pleasant-looking bar wench stepped out from behind the bar, greeted them in a language neither of them understood and set a bottle with two tumblers in front of them.

Aideen whispered to Benji, "I'm not sure what she said and I don't want a glass of this fizzy-looking wine." She beckoned to the barmaid and asked, in what she hoped might be interpreted as Spanish, "Would you have two bowls of pottage and ale?"

The barmaid responded in a confused mixture of English and Portuguese, "Yes, we have a room, but it only has one bed. Where will your son sleep?"

Aideen looked at Benji and suppressed an embarrassed smile. "Why not? It's sure nobody here knows us and we've been too many days with no privacy." She stood up, grasped his arm and headed upstairs to the room. The flickering light of two candle sconces cast an amber glow. Benji immediately flopped onto the cot and tossed the bolster at Aideen like a playful boy. "So. I'm your son, eh?"

Aideen sat near him at the head of the cot. "How old are you anyway? You've never mentioned your age. Shouldn't I know?"

"You know, I'm not sure. I haven't kept track. Does age matter?"

"No. It doesn't matter at all." She edged closer to him. "Not since you have traveled the world, overcome Viking cruelty and learned to talk kindly with the women you encounter."

He lifted a hand to the side of her head and stroked a lock of her long hair.

"Yes, but remember, there's to be no talk in here of fish or ships."

***

When they walked back down the stairs, Aideen felt the thrum of her heart and knew the glow in her cheeks would give her away to what had transpired in the room above. She took Benji's hand, ignoring the barmaid's cool-eyed stare. Searching the room for de Portu, she spotted him surrounded by a group of fishing captains. Bottles of Portuguese *vinho verde* covered their table and the basket of her salted fish was practically empty. Benji pulled two chairs up to the long table and he and Aideen joined the discussions.

One Portuguese fisherman was speaking for the group. "My name is Gregório Esteves. These men call me Goyo. Some of us have heard the fantastic tales of land beyond the horizon, but to be honest, there has never been a captain who dared venture far enough to confirm it. But our question to you, if such tales are true, why would a Basque ever volunteer such valuable information to us? We are competitors."

De Portu was emphatic. "Why? In return for your salt. I will be more than rewarded if you allow me access so I can distribute it to my fellow Basque fishermen who will fish alongside you."

"You have seen how much salt we have in our *salinas* to be sure, but can you guarantee the fish will be there in your Notre Dame Bay?"

Benji spread Niels' chart onto the table. "They say 'seeing is believing', Goyo. And if you follow this chart I guarantee you will scarcely believe what you see."

The eager fishermen poured over the map, fascinated by how their own latitude of Averio was the same as this fishing ground called Notre Dame Bay. Benji assured them he had been born in Greenland and had already fished in Vinland twice.

"Since you assure us you have seen it," Goyo conceded, "perhaps we may be more interested, particularly when we see these large fat fish, but how many days are required to make the crossing from Averio to this Notre Dame Bay as marked on this chart? How can a voyage of such distance be possible?"

"This voyage is not for most men," de Portu bragged, "but it has not been too far for us Basques. When we arrive home, we fill people's stomachs and our purses. You could do the same for yourselves and the people here in Averio."

"*Todos,*" interrupted Aideen, trying her best to use one of the few words in Portuguese she had picked up. "Every fisherman here and all your communities will be fed, Captain Goyo. The fish in Notre Dame Bay can even feed all your country and many more. You just need sufficient rations for each man on the long trip and certainly a few kegs of cider to ward off the scurvy."

Aideen rapped her wine tumbler on the table. "One last thing, gentlemen. I made a promise to the people of Dingle that my efforts would provide good seafood for every soul in County Kerry. To keep that promise I need each of you to stop at my warehouse on your return from Vinland and deposit a fair portion of your catch as my share."

She held her breath and looked from one fisherman to another. She needed their agreement. One by one, after long, silent consideration and some murmured conversations, they began to nod in agreement. Goyo nodded directly at her.

A smile spread across Aideen's face, and she slid her hand over to squeeze Benji's. She had kept the promise and now looked forward to her future.

The End

After a career as a marine biologist in the Maritime Provinces and delivering hundreds of commentaries on the fishery for CBC radio, it's no wonder the sea is prominent in my work, including four non-fiction publications on coastal resources. I am now retired and live on the water near Halifax, Nova Scotia, giving voice to the personalities who pioneered the harvest of our marine resources and opened up this continent. *Aideen's Promise* is my first novel.

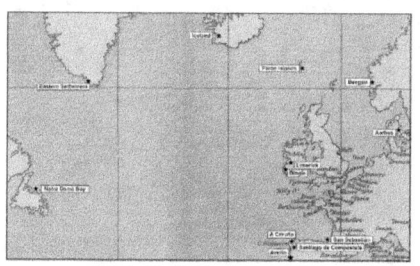

www.ingramcontent.com/pod-product-compliance
Lightning Source LLC
Chambersburg PA
CBHW051416290426
44109CB00016B/1324